Church and Mission in a warming world

Claire Dawson | Dr Mick Pope

© Claire Dawson and Mick Pope, 2014

Published 2014 by UNOH Publishing

2/6–12 Airlie Avenue, Dandenong
Victoria 3175, Australia
www.unoh.org/publishing

All rights reserved. No part of this publication may be reproduced, stored in a retrieval system, or transmitted in any form or by any means – electronic, mechanical, photocopy, recording, or any other – except for brief quotations in printed reviews, without prior permission of the publisher.

Scripture verses from Sections 1 and 2 are taken from the New Revised Standard Version of the Bible, copyright © 1989 by the Division of Christian Education of the National Council of the Churches of Christ in the United States of America. Scripture verses from Sections 3, 4 and 5 are taken from the Holy Bible, New International Version®, NIV®. Copyright ©1973, 1978, 1984, 2011 by Biblica, Inc.™
Used by permission of Zondervan. All rights reserved.
www.zondervan.com The "NIV" and "New International Version" are trademarks registered in the United States Patent and Trademark Office by Biblica, Inc.™

National Library of Australia Cataloguing-in-Publication entry:

Creator: Dawson, Claire, author.

Title: A climate of hope: church and mission in a warming world
Claire Dawson and Dr Mick Pope.

ISBN: 9780994202321 (paperback)

Subjects: Climatic changes--Religious aspects--Christianity. Global warming--Religious aspects--Christianity. Environmental responsibility--Religious aspects--Christianity.

Other Creators/Contributors: Pope, Mick (author), McKenna, Jarrod (foreword), Blair, Sharmila (epilogue)

Dewey Number: 241.691

Authors: Claire Dawson and Dr Mick Pope
Design: Les Colston at Urban Zeal
Editors: Darren Cronshaw and Gabriel Hingley

Cartoons by Ray Higgs

With this book Claire Dawson and Mick Pope have set deep concern alongside shining hope, as they issue a timely call to action for Christians to get serious about climate change. This is urgent business, and I hope and pray that this work will ignite a righteous energy to care for creation.
Reverend Tim Costello AO, CEO World Vision Australia

What a terrific title and book! In these days of climate despair, what a gift to have an accessible book that gives grounds for *A Climate of Hope*. Not "pollyanna-ish", positive thinking but Bible-based, scientific, prayerful, practical, realistic, radical hope. These two deeply committed reflective practitioners and lay eco-theologians, a male meteorologist and female environmental accountant, complement each other seamlessly. Both are leaders in Australian evangelical environmental engagement through Ethos Environment and other organisations. This is a book that challenges but does not cripple; it enables principled, prayerful, practical action. The message is bold but balanced, hopeful but lamenting the likely loss of so much of God's glorious creation. Don't buy it if you don't want to be challenged or changed.
Rev'd Dr Gordon Preece, Director of Ethos: EA Centre for Christianity & Society, Chair, Melbourne Anglican Diocese Social Responsibilities Committee

It was a pleasure to read Claire Dawson and Mick Pope's much-needed and well-researched book. With exceptional clarity, the authors elucidate their broader understanding of Scripture, the basic science of climate change, the politico-economic context in which the problem has arisen and the nature of Church responses to date. They patiently answer various hesitations Christians have expressed around these matters. Their tone is conversational, compassionate and positive, but their understanding of the issues is sophisticated. Claire and Mick end by challenging the reader with inspirational stories and practical suggestions for taking action. I have no hesitation in recommending this book to anyone who is dubious or who has not yet become fully engaged with these issues. It is brilliant!
Thea Ormerod - President, Australian Religious Response to Climate Change

A Climate of Hope is an urgently needed resource for Christians in Australia. With great expertise and care the authors explain the science of climate change and provide a very helpful theological framework in which to understand and respond to this immense ecological and moral crisis. Importantly, you don't need to be a scientist or theologian to read this book, but you do need to be willing to respond to its call to faithful, biblically informed action.
Steve Bradbury - Director Micah 6.8 Centre

 **Claire Dawson** studied Commerce at Monash University before completing a Master of Divinity at the Bible College of Victoria (now MST). She is passionate about connecting the dots between faith and everyday issues, including climate change and consumerism. Claire is married with two young children.

 **Mick Pope** has a PhD in meteorology from Monash University and undergraduate studies in theology from Tyndale College, Sydney. He writes and speaks on climate change, eco-theology and the relationship between science and Christianity. Mick is married and has a son.

Dedication

This book is for Jake (age 12), Sarah (age 5) and Micah (age 2)

– in the hope that you will inherit a habitable and flourishing world.

MP & CD

Acknowledgments

Claire thanks family and friends for their support, encouragement and prayer, including members of the Langwarrin Vineyard Church community. Particular thanks go to husband JD, and the kids, for their patience!

Mick thanks his family, Rev Dr Gordon Preece and Rev Dr B Ward Powers.

Thanks go to Darren Cronshaw and Gabriel Hingley at UNOH Publishing for their hard work as champions, encouragers and editors. Thanks also to Gabriel Hingley for permission to use his own artwork on the cover.

Our sincere appreciation is also extended to Byron Smith, Steve Bradbury and Ben Thurley for their helpful comments in response to our initial draft. Huge thanks also to Nils von Kalm for his careful proof reading of our final draft.

Les Colston at Urban Zeal deserves high praise for an amazing job making this book look so good – better than we could have asked or imagined!

We are particularly grateful for Sharmila Blair's involvement: for being willing to 'road test' this book as a reader and for sharing her journey as a fitting epilogue to the book.

We thank Jarrod McKenna for his willingness to contribute such a heart-felt Foreword for our book. We are inspired by his life, his commitment to justice-seeking, and his ongoing advocacy for stronger action on climate change.

There were a significant number of people who made time to share their stories of hope, or to point us in the direction of helpful people and resources. Collectively these stories are what has embedded significant hope in this writing project. Thanks in particular to Thea Ormerod (ARRCC), Jan Down & Sally Shaw (Transition Town initiatives in VIC and SA), Dave Tims (UNOH NZ) and his Randwick Park neighbour Raymond Diaz, Geoff and Sherry Maddock (Kentucky, USA), David Lewis (CBM), Nils von Kalm (Anglican Overseas Aid), Mike Penberthy (TEAR), John Altmann (Barefoot

Power), Sally Quinn (Green Collect), Cath James (Environmental Project Officer, Uniting Church Vic/Tas), Geoff Westlake (Cheers Neighbourhood Network, WA), Lowell Bliss (Eden Vigil), Russ Pierson (GreenFaith Fellow, USA), Liellie McLaughlan (SA), Bron Adderley (VIC), and Vicky Balabanski (Flinders University, SA). To everyone who has been a part of this journey we offer our sincere thanks - for your time, encouragement, inspiration and support.

Finally, we owe a deep debt of gratitude to all of those who have chipped away at this incredibly important issue for decades, often without much thanks or respect. We may not have met you, and we may not have seen your good works. Yet we believe that God sees and knows, and that your acts to conserve creation are most precious in his sight.

While there have been many people involved in bringing this book to life, we are aware that this book is in many ways far from perfect. Thanks for reading it anyway, and for extending us some grace as you do. We take complete responsibility for any errors and omissions.

Tribute

We wish to acknowledge the precious life and work of Ross Langmead, a true saint and pioneer when it comes to faith, justice and creation care. We had looked forward to Ross' involvement in this project; however his days were cut short so unexpectedly. We thank God for Ross and the rich legacy that he has left. We pray that we might in some small way continue his valuable work.

EVERGREEN GOD
(by Ross Langmead)

May you live in the Spirit of the evergreen God,
In the sun and the showers of blessing.
May you draw your life from the flow of God within,
From all the tender times you bring each other grace.
And for love and for energy be thankful.

May your roots go deep in the warm, dark soil
In the heat and the drought of the summer.
May you stand and trust, knowing God is in it too.
The seed lies in the dark, the night before the day.
Letting go in your growing, still be thankful.

May you grow in the beauty of a colour-filled life
In the joy, in the flowers of creation.
When you sing, when you paint, when you laugh
and when you taste,
May all your senses know that you create with God.
Let the artist within you now be thankful.

May your leaves give shade to the creatures of God
And may birds find the shelter of justice.
May your life together be open to the world,
Embodying the love that springs up from the deep
And transforms all who feel it and are thankful.

Contents

Foreword   13

Section One: How green is my Bible?

Introduction First things first   21
1.1 "Trust me, I'm a scientist": worldviews and who we listen to   23
1.2 In the beginning: why God is green   27
1.3 Who's the Boss? If God is in charge, can the climate change?   35
1.4 Being a disciple of Jesus involves more than saving souls   41
1.5 Resisting empire: politics, religion and caring for creation   46
1.6 More than Good Samaritans: the ethics of climate change   54
1.7 It's the end of the world as we know it (and I feel fine): Does the world have a future?   60
Conclusion Climate Change: a consequence of sin   67

Section Two: The nature of science

2.1 Science works!   71
2.2 Some like it hot   75
2.3 Clear and present danger   83
2.4 Beyond reasonable doubt   89
2.5 Gaze into my crystal ball   94

Section Three: Understanding the times

3.1 Cool, calm and collected   103
3.2 Prophecy & Pride   108
3.3 Affluenza and "Gross" Domestic Product   124
3.4 "Biggering and biggering": market and media   132
3.5 Polluted Politics   140
3.6 A Complacent and Complicit Church   155
3.7 A Called, Compelled and Courageous Church   167

Section Four: Stories of hope

Introduction Fuel your imagination — 179
4.1 Grassroots involvement — 181
4.2 Alongside the poor — 194
4.3 The triple-bottom line: Profit, People, Planet (3BL) — 202
4.4 The Aussie Church in action — 208
4.5 Alongside the church — 216
4.6 Inspiration has a face and a name — 229
Conclusion Next steps together in a season of opportunity — 234

Section Five: What on Earth do we do now?

Introduction The call to live in "climate truth" — 239
5.1 Godly Sorrow — 242
5.2 New ways — 250
5.3 Getting started — 253
5.4 The modified mantra: Resist, Reduce, Reuse, Recycle, Repair — 262
5.5 Other avenues for change — 270
Conclusion Start by taking one step at a time — 283

Epilogue — 285
More information? — 289

Foreword

We're not scaremongering

This is really happening, happening

We're not scaremongering

This is really happening, happening

-Radiohead, *Idioteque*

You, dear reader - if you are like me - have been waiting for the book you hold in your hand.

Why we have been waiting is harder to articulate without being vulnerable. Even biographical.

So here's my personal attempt at describing why *I've* been waiting.

1. The love of my family.

2. The silent faithfulness of the stars.

These are the same reasons I would give in answer to both the question, "What led you to Christ?" and "Why do you care so passionately about creation?" Like most toddlers, my experiments with civil disobedience started early. I refused to go to bed, let alone go to sleep, until my dad or my mum would take me up in their arms to stand underneath the night sky to say goodnight.

Not goodnight to mum and dad. That could wait and find its place in the liturgy of tuck-ins, prayers and sincere questions, giggles, stories and other general delaying tactics.

No. I insisted on being outside in the arms of a parent to say good night to the stars.

"Nigh' nigh' moon. Nigh' nigh' stars."

If raining it would happen under the patio. If cloudy it would happen accompanied by lessons of what is there, even though we can't see it. But normally I was held in the garden underneath a large gum tree.

"Nigh' nigh' moon. Nigh' nigh' stars."

My parents are still bewildered by it. How does that become a child's 'thing'?

I never wanted to be an astronomer. Nor an astronaut. I never wanted to travel elsewhere. It was the night sky that taught me to be *here*. That I belong here. That I'm a part of this dance that is God's good creation. As a little child I was aware that my cries, my prayers and Sunday school songs joined *everything* in worship. Kookaburras and plankton. Giraffes and whales. Old growth forests and frogs. Everything cries out for salvation and sings in worship.

It was under these ancient lights, in the safety of my parents arms, where I learnt to wonder.

The Hebrew word for wonder is *yare'*. Rabbis, preachers, pastors, priests and your garden variety exegete have been left with a choice of how to translate into English. Most Bibles translate it "fear". Given our ecological crisis, "fear" is more than ever understandable. And reasonable. Yet I invite you to stand underneath the stars and interpret *yare'*.

In the loving arms of my parents, gazing at the beauty of the stars, I learnt that *wonder* is the beginning of wisdom. I learnt that God is wonder-full. I learnt to be in wonder of *GOD*.

Choosing *wonder* over *fear* -as a way of engaging with Holy Scripture or our world- is not without challenges.

Like the Bible stories that I had read to me before bed, my story as a child moved from the garden and loving embrace into the cool of the night. Unlike Cain I didn't kill my sibling, although it's reported I came close on a couple of occasions ... (forgive me Elisha!). Year 1 was the first time I became aware of being schooled in this 'Babel project'. The classroom quickly taught us to forget that we too were a part of God's good creation. In a Babylonian education system, they unintentionally sought to instruct us out of our 'creatureliness'.

There was no time in the curriculum for wonder. You need to sit at your desk, inside. You have to wait till recess to go climb. Or run. Or go to the loo. Or sing. (All things I apparently struggled with). Even beauty and creativity was made competitive and "productive". Even the words I learnt for the other-than-human parts of creation secretly silenced my imagination: "the environment" and "nature" became "out there"- making what we are a part of, and all depend on, completely "other".

We are trained early that growth is society's supreme virtue. After all, unlike the garden, empires have towers to build. God made us to be icons, yet empires form us to be slaves, with the building blocks of this civilization, like those of ancient Egypt, made by slaves in sweatshops that are kept out of sight. Even when I was learning about colonisation, this was a story rarely told as victors hide the bodies under flags. The cries of the land, like the cries of the oppressed, are made inaudible over the noise of "business as usual". Most teachers - some yelling, some kindly taking me aside - told me to "Stop daydreaming!" I was in high school when I realised I was in exile from what God was and is acting to redeem... creation.

And then, in my lostness, Jesus. Francis of Assisi's conversion through kissing a leper is known to many. Fewer know that the journey of Francis' transformation began as he stumbled drunk out of a pub. This returned solider, drinking away the trauma of wielding a sword and not a ploughshare, looked up at the night sky and was struck with wonder: "If these are the creatures, how beautiful must the Creator be."

I'm no saint. Yet I too, was led to Christ by the silent contemplation of the stars.

Nearly as soon as I responded to the good news that, "God so loved all of creation that he sent his Son..." I'd hear preachers say things that contradicted the testimonies of the stars.

I hadn't read the theology of the early church fathers, but bobbing up and down with a board between my legs and the spray of breaking sets on my back, the waves at Trigg Beach witnessed to something different to the fiction of being "Left Behind". I hadn't read great Biblical scholars like N.T. Wright, but reading the Psalms in the bush before school clashed with talk at funerals of "going home." I hadn't read the Magisterial Reformers, so I was unaware that Calvin had written what the stars whispered, that "everything in [creation] tells us of God" and Luther's words, "God writes the Gospel, not in the Bible alone, but also on trees, and in the flowers and clouds and stars."

I knew deeply then, that God's presence is my home. But as I sang on the way to school to the God who I learnt loved me, and learned to hear the praise of bottle-brush bushes and grass-trees , I knew that presence wanted to flood our home, not destroy it. During teenage Bible studies on being "aliens and exiles", I was certain that I wanted to be alien to an empire which destroys the creation that Jesus came to save.

Look up from your page. If inside, go to a window. Where you are is where hope is coming. The New Jerusalem descends, here. The New Creation is the transfiguration of – not the doing away with - what you are looking at now. By grace, through Jesus, Heaven is coming here.

So in a way being 'left behind' sounds rather lovely. Especially if Jesus is coming back.

And it turns out, that's exactly what all orthodox Christians have believed throughout the ages.

So if your journey is anything like mine, welcome home. This book will be a friend and a resource. God is calling a generation of eco-prophets to announce -with Calvary-like nonviolence- that the Gospel is good news to our unprecedented ecological crisis. To this you are called, at such a time as this. Until that time when God's presence covers the earth like the water covers the sea, may you learn the kind of wonder Father Zossima talks of on his death-bed in Dostoyevsky's *The Brothers Karamazov*:

> "Love all God's creation, the whole and every grain of sand in it. Love every leaf, every ray of God's light. Love the animals, love the plants, love everything. If you love everything, you will perceive the divine mystery in things. Once you perceive it, you will begin to comprehend it better every day. And you will come at last to love the whole world with an all-embracing love."

… just as our Lord does.

Jarrod McKenna

Co-Founder of First Home Project, Pastor and National Director at Common Grace

[Monday of the third Week of Ordinary Time, Nov. 2014]

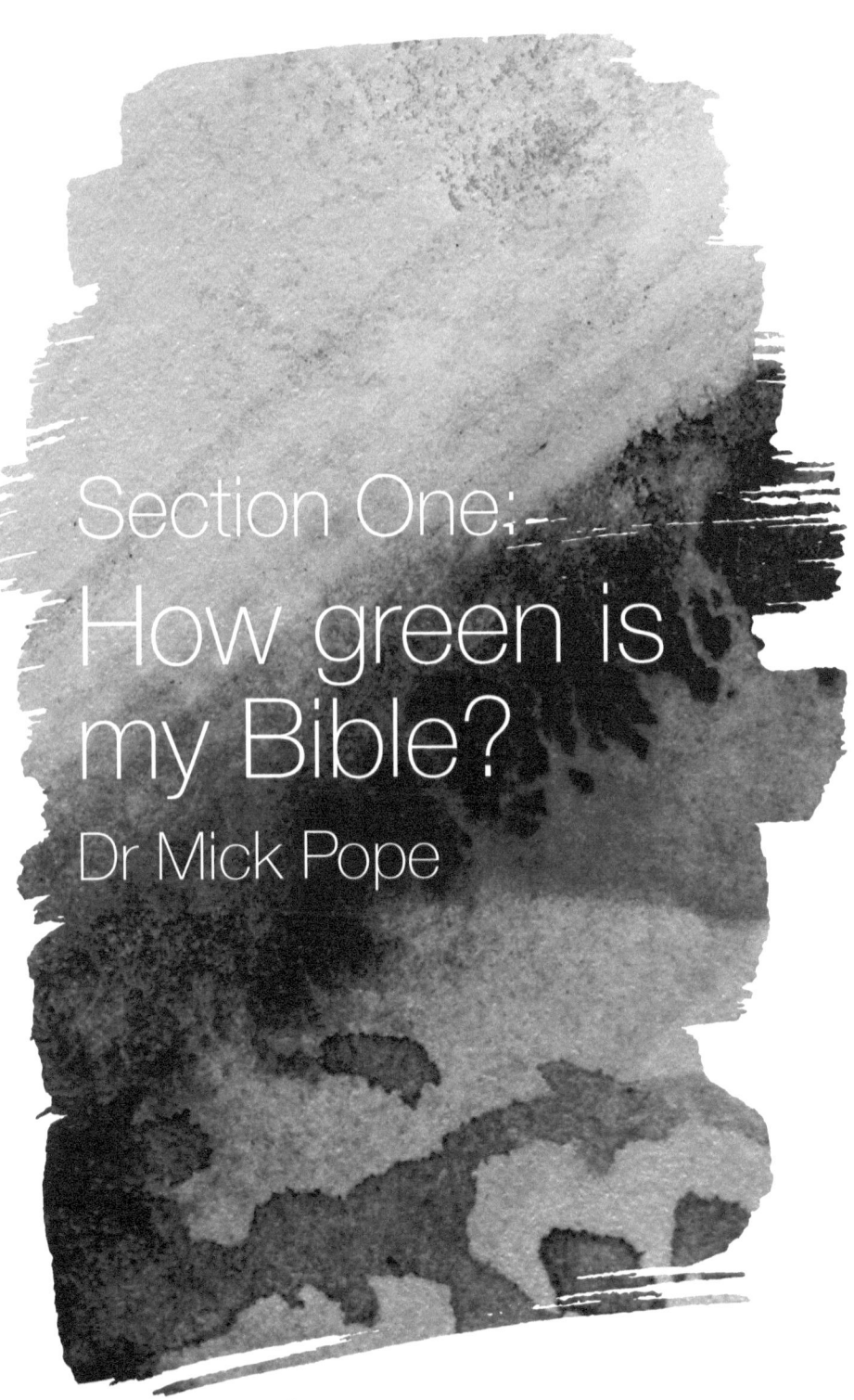

Section One: How green is my Bible?
Dr Mick Pope

Introduction
First things first

When we started to talk about writing this book, Claire and I thought that it would be helpful to start with the Bible. Maybe you have concerns that accepting the idea of human-caused climate change is somehow unbiblical. Or maybe you're convinced climate change is happening and you want a biblical framework through which to understand why it is happening, and what to do about it. Perhaps you are not a Christian, and a friend has thrust this into your hands to try and show you that Christianity isn't opposed to action on climate change. If this is you, while I have tried to limit the religious jargon and define things as I go, you may find this section a little bit of work. In any case, if you'd rather start with the science, feel free to jump ahead to Section Two and come back to this later.

Either way, it is important to think through how the Bible addresses the sorts of issues that climate change raises. This is because the Bible is the foundational document for Christian belief. For Christians, we believe the Bible is a collection of inspired writings covering the history of the people of Israel and the early Christian church. Even if you don't believe the Bible is inspired by God, knowledge of what it says will enable you to see that being a Christian is not incompatible with a concern about climate change, despite all too many examples to the contrary!

This is how this section will proceed. Firstly, we need to think about how the world works, and who we can trust. Does it matter if a climate scientist is a Christian or an atheist? Can we trust what they say when some prominent Christians don't? And what should a Christian have to do with science anyway?

Once we've done with that, we'll start to think about creation. We won't get tied up in questions of how it was done but instead look at issues such as 'what is creation for?', 'what is our place in it?' and 'does the non-human creation matter to God?'[1] Working out God's relationship to creation leads into a discussion about how he rules it, and whether or not humans can actually change the Earth's climate in radical ways apart from God's permission, if we can change it at all.

Once we've unpacked how God relates to creation and how things are meant to work, we need to then address the question, 'what does this have to do with Christians? Isn't our mission to save souls? Aren't the keys to the Christian life Bible reading, prayer, evangelism?' Taking a deeper look will show us that the Gospel Jesus preached was not simply about 'going to heaven when you die'. The Gospel is about the whole world being put right –the broken relationships between people and God, people among themselves, people within themselves, and people with the rest of creation. That Jesus was *raised from the dead* points to the fact that matter itself matters to God. The whole world has a future, as described for example in Revelation 20 and 21.

Sometimes one of the best ways to love your neighbour is to love the Earth you share with your neighbour. When you put all of this together, caring for people and caring for creation are all Gospel acts. Doing these things opens up opportunities for 'Gospel conversations' – explaining what motivates us.

This idea of loving our neighbour by caring for the creation and avoiding the worst of climate change leads us to look at the parable of the Good Samaritan, *the* story about neighbours. However, we can't properly appreciate this parable from every angle without looking at the back story of how the Bible confronts empire, and that to declare that Jesus is Lord means that Caesar, and all of his modern successors, is not. This is the thorny issue of religion and politics.

---

1 What I want to avoid is any debate over creation versus evolution, or the age of the Earth. I have known both creationists and those believing in a form of divinely guided evolution to be concerned about climate change. Likewise, while a young Earth view makes it harder to interpret the record of climate into the ancient past, there is plenty of evidence over the past few hundred years (if not thousands) to show that humans have had a discernible impact upon the climate system.

Finally, if it is all going to be destroyed, then who really cares? That said, does the Bible really say the Earth is going to be destroyed in judgement? Once I've shown it doesn't actually indicate this, the last objection to Christians combating climate change vanishes.

So let's begin our exploration...

1.1 "Trust me, I'm a scientist": worldviews and who we listen to

Key points:

Everyone has a worldview, consisting of stories, questions, symbols and ways of acting.

Christians shouldn't reject the views of climate scientists simply because some of them don't have a Christian worldview.

A Christian worldview doesn't have to be anti-science or suspicious of climate change science.

You may already have an opinion about whether or not human beings are changing the climate. Regardless of the view you take, it's likely that you didn't arrive at that opinion based purely on the scientific evidence. We all have worldviews and they filter what we see and hear.

This isn't to say that there are no such things as facts, but rather that all facts come to us through our worldviews. A worldview consists of four things: stories we tell ourselves, questions these stories answer, symbols that define our identity and things that we do to act out our stories.[2]

---

[2] For a discussion of worldviews, see for example N T Wright, *The New Testament and the People of God* (Minnealopis: Fortress Press, 1992), 38. I tease this out in reference to our views on environment and the idea of eco-mission or the mission of the church to creation in Mick Pope, "Preaching to the Birds? The Mission of the Church to Creation," in *Speaking of Mission Volume 2*, ed. Mike Frost (Morling College: Morling College Press, 2013).

For a Christian, the over-arching story we tell ourselves is the Bible story, yet there are some differing ideas regarding exactly what this is. How you understand the central themes of the Bible will affect how you understand ideas like climate change. Here I will examine the way in which I understand this story, how it affects how I understand climate change, and how I think we should be responding.

Every scientist has their own worldview, which can be quite different from a Christian one. However, there are many climate scientists who are Christian, including a former Head of the Intergovernmental Panel on Climate Change, Sir John Houghton. So can we trust them when they say that the climate is changing and that us humans are largely responsible?

The word 'science' comes from the Latin word *scientia* which means 'knowledge'. It is a process of observing, theorising, making hypotheses and testing them. Ideally, science changes over time in response to new ideas, new data and pushing theories until they break and need modifying or abandoning for something better.

Scientific theories are attempts to describe the world in the most concise manner possible. The laws of physics, which climate science uses, are shorthand for the regularity we observe in nature: the repeatable patterns and behaviour of physical systems. Of course, lots of things we see are highly complex, which make predictions difficult. That said, science has been very successful in describing how the world around us operates. This can be seen, for example, in how the increasingly detailed sets of observations of climate or increasingly sophisticated climate models have given us greater confidence that humans are changing the climate.

It is true that a scientist's worldview can affect their science. Recognised or not, scientists operate with a philosophy that shapes the kind of hypotheses they can make. Albert Einstein did early work towards the discovery of Quantum Mechanics, the ideas that describe the spooky world of the very small. Yet as he grew older, his view that 'God does not play dice' meant he couldn't accept the consequences of Quantum Mechanics.[3] Physicist Fred Hoyle's atheism didn't allow him to accept the idea of the Big Bang because it smacked too much of Genesis 1 and a role for a Creator, yet even he couldn't entirely resist the idea that the universe appeared designed.[4] More recently, scientists like Richard Dawkins have insisted that evolution renders God highly improbable.[5]

So, can Christians trust scientists to tell us the truth? Should we automatically be suspicious? In Genesis 2, God invited Adam to name the animals, which implies that humans were created to classify and understand the world around us. The Bible also insists that the reason things so often go wrong in human society is that human beings are sinful. Sin (rather than 'sins' in the plural, which are individual immoral acts) can be described as missing the mark, or falling short of what it means to be human, which is to reflect God (Romans 3:23). While sin may affect us morally so that we can't think correctly about spiritual things without God's Spirit (Romans 1; 1 Corinthians 1), it doesn't mean that a person needs the Spirit to think about the world around us.

Two examples from the Bible make this very clear. The Bible is understood by Christians to have been inspired by God. The authors of the Old Testament were Jews, God's people. And yet King Lemuel, who wrote some of Proverbs, may have been a non-Jewish king. Other proverbs echo Egyptian wisdom literature, so it isn't like Jews or Christians have a monopoly on common sense. That said, there is no shortage of Christians involved in climate science. Katherine Hayhoe (author of the book *A climate for change*)[6] and the aforementioned Sir John Houghton are just two examples.

---

3 See for example William Hermanns, *Einstein and the Poet: In Search of the Cosmic Man* (Brookline Village: Branden Press, 1983), 58.

4 See Robert John Russell, *Cosmology, Evolution, and Resurrection Hope: Theology and Science in Creative Mutual Interaction* (Kitchener: Pandora Press, 2006).

5 Richard Dawkins, *The God Delusion* (Boston: Houghton Mifflin Harcourt, 2006), 176.

6 Katharine Hayhoe, *A climate of change* (Nashville: Faithwords, 2009).

Augustine made this point well in his book *The Literal Meaning of Genesis*:

> Usually, even a non-Christian knows something about the Earth, the heavens, and the other elements of this world, about the motion and orbit of the stars and even their size and relative positions, about the predictable eclipses of the sun and moon, the cycles of the years and the seasons, about the kinds of animals, shrubs, stones, and so forth, and this knowledge he holds to as being certain from reason and experience. Now, it is a disgraceful and dangerous thing for an infidel to hear a Christian, presumably giving the meaning of Holy Scripture, talking nonsense on these topics; and we should take all means to prevent such an embarrassing situation, in which people show up vast ignorance in a Christian and laugh it to scorn.[7]

What all of this makes clear is that there is much in the world that is common knowledge, facts about the world that God makes clear to anyone with sufficient technical understanding. This natural knowledge should come as no great surprise from a God who also exercises common grace: God does indeed send rain on the righteous and unrighteous (Matthew 5:45). God is a generous God, and the world is full of things for anyone to discover!

The sum of all this is, even if someone has a worldview very different from ours, they may still understand something better than us because of their knowledge and training. You don't get your hairdresser rather than your doctor to diagnose your illness simply because one is a Christian and the other isn't, so why do so many trust pastors, bloggers, politicians and the media over climate scientists? I'll deal with the scientific evidence in Section Two, but for now, let's turn to the Bible to try and understand more about the creation and our role in it.

---

7 St Augustine, "The literal meaning of Genesis," in *Ancient Christian Writers Volume 1*, John Hammond Taylor (New York: Paulist Press, 1982), chapter 19.

1.2 In the beginning: why God is green

Key points:

The Bible contains a truly green theology.

Genesis 1 describes creation as a temple which means it is sacred to God. Looking after creation is service to God.

Since all humans are made in God's image, caring for creation is a matter of justice, as a changing climate brings harm to others.

Psalm 104 tells us that God cares for all of creation, not just what is important for human needs; hence we need to care for it as well.

Even though Jesus tells us we matter more to God than sparrows, all living things matter to God.

You may worry that getting involved in environmental issues such as climate change may lead you to compromise your faith in some way. Will you be swayed by green, atheist or leftist ideology? There certainly are some in the environmental movement who might be labelled as nature-worshipping in some sense, and some even relativise or denigrate the role of human beings in nature.

We should admit though, that for too long, large sections of the Church have abdicated their responsibility to properly care for creation. In our efforts to avoid viewing nature as divine, we have seen it as disposable, and reaped the rewards of this view in the form of environmental destruction (which includes climate change).

We have also earned the disdain of many as we have generally stared at environmentalists across a very wide divide. But need we always view environmentalists with suspicion? And isn't there a solid theology of ecology at the heart of the biblical faith? We believe so, and it lies in the fact that Genesis 1 describes *what* creation is for, rather than precisely *how* it was made.

Cosmologies are stories that explain aspects of the world around us. As John Walton notes in his book *The Lost World of Genesis One*, the cosmology of Genesis looks a lot like that of the nations surrounding ancient Israel in some of its details, and yet it is quite different in others.[8] It is not, however, anything like modern scientific cosmologies, and recognising these differences is important.

Theses two ways of thinking may be understood by considering the building of a house. We can talk about the architectural plans, the building materials, how many workers were involved and so on. But a house does not become a home until it is inhabited, decorated and invested with time, care and emotions. While we can talk about how a house is constructed, a home goes well beyond the mechanics of mere building to what rooms are *for* and the family relationships that take place there. So, following Walton on Genesis 1, we should be far more concerned firstly about *who* made the creation (the God of Abraham, Isaac and Jacob, and the Father of Jesus) and about *what* creation is for, what our place in it is and how to care for it, than arguing over precisely *when* and *how* it was made.

One of the things this understanding does is to tie human culture and creation together. Michael Welker, in *Creation and Reality*, notes that creation is not to be simply identified with nature, but includes it.[9] This is evident, for example, in the role that lights in the sky play in marking out days, seasons and years:

> And God said, "Let there be lights in the dome of the sky to separate the day from the night; and let them be for signs and for seasons and for days and years (Genesis 1:14).

---

8 John Walton, *The lost world of Genesis One* (Nottingham: IVP, 2009), 12.
9 Michael Welker, *Creation and Reality* (Minneapolis: Fortress Press, 2009), 21.

The heavens are the place where natural forces determine life and culture; they are not simply the object of the study of astrophysics. More than this, human beings are central to the story of creation, not a distraction from it. We are neither to be entirely excluded from nature, nor to have unfettered use of it.

Interestingly, the first creation account ends with God resting on the seventh day (Genesis 2:2; Exodus 20:11).

> And on the seventh day God finished the work that he had done, and he rested on the seventh day from all the work that he had done (Genesis 2:2).

In the ancient Near East, temples were the place where Gods would rest, but this rest was not the same as being idle. Rather, the god would rule over their people from their temple. Their presence in that temple was represented by an idol or image. We find a very similar idea in Psalm 132, where God's resting place and throne is the Ark of the Covenant, which is kept in the temple in Jerusalem.

> This is my resting place forever; here I will reside, for I have desired it (Psalm 132:14).

Walton concludes that Genesis 1 recounts the establishing of the functions of a cosmic temple from which God can rule. This can be seen in various features of the Jerusalem temple. The water basin represented the sea and the temple pillars the pillars of the Earth (1 Kings 7). The Hebrew word for light used in connection with the temple lamp (Exodus 25:6) is the same used for the lights in the sky created on the fourth day (Genesis 1:14).

> And God said, Let there be lights in the dome of the sky to separate the day from the night; and let them be for signs and for seasons and for days and years (Genesis 1:14).

So God rules from his cosmic temple, and it is here again that we see the important role given to humanity – to serve God as his representatives. Rikk Watts, in an essay in *What Does it Mean to be Saved?*, notes there are close parallels between the formation of humans in Genesis 2 and how ancient idols were made.[10] Described as a temple, the creation is therefore sacred. However as the creation of God, it is not divine. Our function as the image-bearers of God is to carry out the mission of God to rule over and care for all he has made, as priests in his sacred temple. Caring for creation is an act of worship of the God who made it. Conversely, behaviour that actively works to destroy creation can be seen as blasphemy.

This is in fact a two-way relationship. The Garden provides for human needs, but humans are also to provide for the Garden. Caring for creation – including dealing with climate change – is not only good management of resources but is also the fulfilment of our proper role upon the Earth. Conversely, to ignore climate change is bad resource management, but more than that, it is unjust.

In the ancient Near East, only the king was said to bear the image of their patron god. Genesis 1 states that all humans bear that image and in Adam we are to be fruitful and multiply (verses 26-28).

> Then God said, "Let us make humankind in our image, according to our likeness; and let them have dominion over the fish of the sea, and over the birds of the air, and over the cattle, and over all the wild animals of the earth, and over every creeping thing that creeps upon the earth." So God created humankind in his image, in the image of God he created them; male and female he created them. God blessed them, and God said to them, "Be fruitful and multiply, and fill the earth and subdue it; and have dominion over the fish of the sea and over the birds of the air and over every living thing that moves upon the earth."

---

10 Rikk Watts, "The New Exodus/New Creational Restoration of the Image of God: A Biblical-Theological Perspective on Salvation ," in *What does it mean to be saved?: Broadening Evangelical horizons of salvation*, ed. John G. Stackhouse Jnr (Minneapolis: Baker Academic, 2002).

Made in his image: the foundation for human rights

Our shared image is the basis of the biblical command not to murder (Genesis 9:6), for example. Yet it is hard to experience the divine blessing spoken of in Genesis 1 when you don't have access to fresh water, food and to other basic needs like health care and education. Climate change threatens all of these things, making it a matter not of care for creation alone, but of justice and fundamental human rights.

The idea of human rights can be somewhat problematic, unless it is underpinned with a fundamental understanding of what it means to be human. Genesis 1 should alone be sufficient, for all people bear the image of God. For the non-religious, our shared evolutionary history is often cited, as well as our shared inhabiting of the world. But from where can we derive a sense of redistributive justice? Psalm 82 for example, makes it clear that God has a concern for the poor, the weak, the orphan, the needy, the lowly and destitute (verses 3-4). They are often in that situation because of the wicked (verse 4) and yet some are so quick to assume that people are poor because they deserve it. Not so hasty! Because God owns all nations, he will judge such things (verse 8). So the Bible gives us a foundation for such fundamental human rights, and the right of people to be protected from the wickedness of human-induced climate change.

One of the blessings of Genesis 1, and a hot topic in the climate change debate, is that of population. Genesis 1:26-28 tells us that we are to be fruitful and multiply and fill the Earth. A number of Christians object to forms of birth control or family planning as going against this command. It is worth thinking a little more closely about this.

Firstly, we have a narrative that is set "in the beginning" (Gen 1:1) – this is pictured to be empty of humans. It makes sense to speak of filling when something is empty, and this exhortation to "fill the Earth" is repeated again to Noah and his family after the flood (Genesis 9:1). But what about when it is full? Could we not say that the Earth is already full? How many humans are too many for the Earth to support?

Secondly, this statement about fruitfulness and multiplying is part of a blessing, not a command as such. What happens in the instance that there are so many humans that we are no longer fruitful; when more mouths to feed is no longer a blessing?

Thirdly, in verse 22, God gives the same blessing to the animals. Do these two blessings need to compete? What happens when they do?

Finally, we need to acknowledge that population is not the only driver of environmental or creation harm. There is a well-known equation in sustainability circles known as IPAT: impact = population x (level of) affluence x technology. We can see this for example in the carbon emissions per capita being much higher in a country like Australia or the United States, compared to say Fiji or Tuvalu (Pacific Island nations suffering from sea level rise). Developed nations generally have more advanced technology and higher levels of wealth, and their environmental impact is consistently far greater per capita than more populous, developing countries.

Population is perhaps one of the most difficult issues within the discussion of climate change, particularly when different faith perspectives and worldviews are involved. While it is a subject matter in its own right and we confess we can't do it justice within the parameters of this book, Claire does revisit this thorny issue again toward the end of the book.

Genesis 1:26-28 also says two other things. Part of God's blessing to humans is that they are to have dominion over all creatures and subdue the Earth. Dominion or rule has already been discussed above, and the idea of creation as a temple and us serving as images or idols of God should put this into its proper context – dominion does not mean domination.

But what of subduing? Land is the object of the subduing (and not creatures) and its subduing is closely associated with filling the land with people. Richard Bauckham sees therefore the subduing as associated with agriculture, i.e. you cannot fill the land with people unless you can feed them.[11] It then follows

---

11 Richard Bauckham, *Bible and Ecology: Rediscovering the Community of Creation* (London: Darton, Longman and Todd, 2010), 16-17.

as above that there are limits to this filling and therefore subduing, as we are viewing an account that imagines the beginning of all things.

The creative God: God loves all he has made, and we should too

We've seen that creation is God's sacred temple, and that we have a central role to play in the proper running of this temple: we are to tend and care for it, as well as benefit from its provision. Moreover, since all humans are made in God's image, every human should have a share in this provision. But does the creation have value beyond what it can do for us, and can it flourish beyond our care? Psalm 104 is a statement about the creative genius of God and how he delights in all that he has made. Verses 10-16 state that,

> He sends forth springs in the valleys; They flow between the mountains; They give drink to every beast of the field; The wild donkeys quench their thirst. Beside them the birds of the heavens dwell; They lift up *their* voices among the branches. He waters the mountains from His upper chambers; The Earth is satisfied with the fruit of His works. He causes the grass to grow for the cattle, And vegetation for the labour of man, So that he may bring forth food from the Earth, And wine which makes man's heart glad, So that he may make *his* face glisten with oil, And food which sustains man's heart. The trees of the Lord drink their fill, The cedars of Lebanon which He planted.

Human economic concerns are only a part of the sum total of God's care. Water is a key concern in places like ancient Israel and in our own native Australia. In this instance the Psalmist spends five verses describing how the Earth, trees, birds and wild donkeys are watered (verses 10-13, 14, 16), and only two verses on human economic activity (verses 14-15). Our need for and use of water forms a small part of a huge water cycle, yet people seek to dominate fresh water sources, using them up, contaminating them and wasting them. For example, a number of rivers in the USA no longer reach the sea,[12] and nature and irrigation compete for water in the Murray-Darling basin of eastern Australia. Climate

---

12 For example the Colorado River, see Peter McBride and Jonathan Waterman, *The Colorado River: Flowing through conflict* (Englewood: Westcliffe Publishers, 2010).

change affects water supplies around the world through changing rainfall patterns, altered evaporation patterns, sea level rise and melting glaciers. These changes affect humans and the rest of the creation alike.

While it is contentious, this raises the issue of how natural resources are managed. The idea that people can own water, or air, seems at odds with God's Lordship over all these things, and their provision for all of humanity and all of creation. I am reminded of a line from the Anglican Prayerbook, "to seek the common good and to share with justice the resources of the earth."[13] Is privatisation always a good thing? Air and water are not about profit, they are about provision.

The Psalmist also reminds us that we all have our place in creation and that humans must not edge out non-human creatures. God has made places for animals to live where they can receive the blessing of fruitfulness and multiplication (Genesis 1:22). Fir trees are for storks, the high mountains are for goats, the night is for lions (verse 17-22) and the oceans are for a vast array of living creatures (verses 25-26). Rising temperatures and an increasingly acidic ocean (due to the burning of fossil fuels) threaten this multitude.

It may be confronting to realise that the world around us does not exist solely for our needs. Everything has its place and is cared for by God. In an age before wildlife documentaries and tourism, God expressed his care for animals that people did not eat, including lions that from time to time ate people's livestock. God's providential care extends to sparrows, and hence all creatures with whom we share the Earth and who are also threatened by our actions (Matthew 10:29-31). That Jesus affirms that we are of more value than sparrows, assumes that sparrows *are* of value.[14] To care for non-humans is not unchristian; it is incredibly Christian because it reflects God's care for them and our own office as his image-bearers, who have been entrusted with a duty to care for them too.

---

13 Anglican Church of Australia, *An Australian Prayerbook 1978* (Sydney: Anglican Information Office, 1978).
14 Richard Bauckham, "First Steps to a Theology of Nature, *Evangelical Quarterly*, 58 (1986), 229-244.

1.3 Who's the Boss? If God is in charge, can the climate change?

Key points:

The flood account describes God undoing creation as a judgment on sin.

The flood was also a saving act as God preserved both humans and animals in the Ark. God wants to save all he has made.

God gives us over to the consequences of our idolatry, and this includes environmental degradation.

Empire and the pursuit of wealth always lead to environmental damage.

Environmental damage and climate change are compatible with more than one way of thinking about God's will and rule of the world.

God will judge those who destroy the Earth.

We've now seen that God cares for creation, is glorified by it, and is firmly in charge of it. One objection Christians sometimes have about climate change is that since God is in charge, he wouldn't let it happen, and if he did it is obviously then his will so there is no point in doing anything about it.

In *High Tide*, Mark Lynas records some words from Christians living in Tuvalu that I find sad.[15] Tuvalu is an island nation of coral atolls which lay just a few metres above sea level. As the world warms, sea water expands, ice sheets melt,

---

15 Mark Lynas, High Tide: *How climate crisis is engulfing our planet* (New York: Harper Perennial, 2005), 94, 118.

and consequently sea levels rise. This leaves atolls like Tuvalu under threat of erosion and the loss of fresh water. When Lynas asked two Tuvaluan schoolgirls if they believed the sea level was rising due to climate change, they answered "No. We're Christians. God will protect the island." Another man thinks, "Only the Creator can flood the world" and "I believe in God – I don't believe in scientists." Is it really true that only God can flood the world? Such an understanding is often based on a reading of Genesis 9:11-17

> "I establish my covenant with you, that never again shall all flesh be cut off by the waters of a flood, and never again shall there be a flood to destroy the Earth." God said, "This is the sign of the covenant that I make between me and you and every living creature that is with you, for all future generations: I have set my bow in the clouds, and it shall be a sign of the covenant between me and the Earth. When I bring clouds over the Earth and the bow is seen in the clouds, I will remember my covenant that is between me and you and every living creature of all flesh; and the waters shall never again become a flood to destroy all flesh. When the bow is in the clouds, I will see it and remember the everlasting covenant between God and every living creature of all flesh that is on the Earth." God said to Noah, "This is the sign of the covenant that I have established between me and all flesh that is on the Earth."

So God promised that there would not be a flood to destroy the Earth where all flesh is cut off by waters. But what was the Flood about? Human hearts were full of evil and the Earth was corrupt (Genesis 6) and God decided to destroy it, yet not utterly. In Genesis 7:11 we read that

> In the six-hundredth year of Noah's life, in the second month, on the seventeenth day of the month, on that day all the fountains of the great deep burst forth, and the windows of the heavens were opened.

As John Walton notes, the Bible presents a view of the cosmos that was the same as the rest of the ancient world, with a solid surface or firmament above through which the hail and snow were allowed to fall out of their storehouses (Job 38:32), and waters below the Earth.[16] It is this firmament that separates the waters above

16 John Walton, *The Lost world of Genesis One*, 29.

(rain, etc) and the waters (oceans) and dry land below (Genesis 1:6-10). Hence, the Flood was an act of "uncreation", undoing the order that God had created, and an act of judgment. This universal destruction is of a different nature to any climate change scenario. The very worst of eventual sea level rise scenarios is about 70 metres, which would be enough to radically alter the world and dramatically disrupt human society, but it is certainly no Noah-like Flood.

The Flood is not only an act of "uncreation" and judgement, but also an act of re-creation and salvation. The receding water was an act of re-creation, since it represented a return to the order of Genesis 1. The Flood was an act of salvation because a remnant of humanity and of the animal world was saved in the Ark, and could reoccupy the land once the waters had gone. God then made a covenant never to perform such an act of "uncreation" again with Noah, his family, and the other living creatures. But does this mean that sea level rises due to global warming simply can't occur?

In order to understand how God can be in charge yet humans can cause destruction upon the Earth, we need to place climate change in the sphere not only of science but of ethics. We need to look at how both science and ethics describe the way the world works. The first idea is to see how God works in the ethical sphere in Romans 1. Christians talk an awful lot about sin but often miss the fact that idolatry is the heart of the issue: worshipping created things rather than the Creator. Three times in Romans 1 Paul states that God gives people over to sinful behaviours because of their idolatry. While there is no mention of climate or environment in verses 28-32, there is a long list of sins that relate directly to ethical causes. Harm to the environment occurs when ideas of economic growth run up against natural limits. The West has benefitted from over a century and a half of fossil fuel burning and the use of the developing world as sources of cheap resources, cheap labour and lax environmental codes. This historical debt in terms of emissions and moral culpability can easily be encapsulated by a list that includes (verses 29-30):

unrighteousness, wickedness, greed, evil; full of envy, murder, strife, deceit, malice; they are gossips, slanderers, haters of God, insolent, arrogant, boastful, inventors of evil, disobedient to parents, without understanding, untrustworthy, unloving, unmerciful.

More than this, we see the link between idolatry, greed, empire and ecological collapse in the Old Testament. Michael Northcott notes that there is a close connection between ecological disasters and exile on one hand, and unfaithfulness to the laws and worship of the Lord on the other, in passages like Jeremiah 5:22-28[17]. There was a direct connection between empire building, the pursuit of pagan idols of fertility and the failure keep the Sabbath rest with injustice and ecological crises. Under such circumstances, ecological collapse was an inevitable consequence. Let's look at how this works.

> Do you not fear me? says the LORD; Do you not tremble before me? I placed the sand as a boundary for the sea, a perpetual barrier that it cannot pass; though the waves toss, they cannot prevail, though they roar, they cannot pass over it. But this people has a stubborn and rebellious heart; they have turned aside and gone away. They do not say in their hearts, "Let us fear the LORD our God, who gives the rain in its season, the autumn rain and the spring rain, and keeps for us the weeks appointed for the harvest." Your iniquities have turned these away, and your sins have deprived you of good. For scoundrels are found among my people; they take over the goods of others. Like fowlers they set a trap; they catch human beings. Like a cage full of birds, their houses are full of treachery; therefore they have become great and rich, they have grown fat and sleek. They know no limits in deeds of wickedness; they do not judge with justice the cause of the orphan, to make it prosper, and they do not defend the rights of the needy. (Jeremiah 5:22-28)

Worship of God includes dependency on him for our material needs. We can see this in Israel's wilderness wandering in Exodus 16 for example, where the collection of manna was strictly regulated. The command from Yahweh was to collect only enough for today and enough for the Sabbath. This is continued

[17] Michael Northcott, *A moral climate: the ethics of global warming* (New York: Orbis Books, 2007), 13.

in the Sabbath regulations. Resting meant reliance on God and not our own efforts. Every seventh year, the land was to have a Sabbath rest (Leviticus 25). This not only expressed a trust in the Lord but also gave the land rest. The Sabbath rest allowed the land to lie fallow, which was important in maintaining its agricultural viability.

The constitution of Israel meant that the land was apportioned so that people could feed themselves, with only the Levites not having to grow their own crops (Numbers 18). But Israel desired to be like the other nations and have a king. Samuel warned that this would mean having the best of the produce of the land taken by the king to maintain a standing army (1 Samuel 8). This is indeed what happened under Solomon (1 Kings 4), and things took an altogether more sinister turn under Ahab, who was not above murder in his greed to acquire land (1 Kings 21). This kind of ruthless acquisition seems to be what Isaiah has in mind when he says "Ah, you who join house to house, who add field to field, until there is room for no one but you, and you are left to live alone in the midst of the land!" (Isaiah 5:8). The expansion of empires always seems to result in environmental degradation.

As indicated in Jeremiah 5, another key to the Israel puzzle is its stubborn idolatry. After the conquest of the promised land (see the Book of Joshua for the account), Israel was tempted to follow the local fertility cults. It was a way of having an each way bet on ensuring the rains would come and the land would yield its crops. King Ahab worshipped the Canaanite storm-god Baal, and as a result, God brought drought to Israel (1 Kings 16-17). When the pursuit of fertility becomes an idol, not only is God abandoned but the land is often pushed beyond its limits. This ancient story should stand as a warning to us as we pursue technology as the sole solution to the climate change problem.

So does climate change happen according to God's permission or his prescription? Is it God giving us over to our sin and its consequences because of our idolatry, or God carrying it out so that there is no alternative for us? I think it is the former. But even if climate change is God acting in judgement, this is not some kind of escape clause for Christians not to act. In Isaiah 10, we read of God denouncing Assyria for its violence against Israel. Yet he had chosen Assyria as his rod of anger and judgment against Israel for her idolatry. Hence, it was not the act of invasion itself that was denounced, but the way in which Assyria carried it out, and that they did not have God's purposes in mind. They were doing what God required of them, but not with the intent that God had in mind. God may be the one who judges nations, but he is not the direct source of evil. He holds us responsible for our sins against his creation and those that suffer as a result. A verse often overlooked by Christians is Revelation 11:18, which warns that those who destroy the Earth are themselves at risk of being destroyed:

> And the nations were enraged, and Your wrath came, and the time came for the dead to be judged, and the time to reward Your bond-servants the prophets and the saints and those who fear Your name, the small and the great, and *to destroy those who destroy the Earth.* (emphasis mine)

To go back to the original question then: God's promise not to flood the Earth again was a promise not to undo creation utterly in an act of uncreation. Sea level rise due to warming atmosphere and oceans, and flooding due to changes in rainfall patterns do not constitute the same kind of act of destruction.

Climate change may be a prescriptive act of judgement by God on the various sinful behaviours and idolatry that lead to global warming, or it may simply be God letting us suffer the consequences of our actions (permissive will). Either way, in the same way that Noah built the Ark to save both humans and non-humans, so the return of Christ will rescue both. I'll discuss this a little later in this section.

1.4 Being a disciple of Jesus involves more than saving souls

Key points:

Jesus calls us to make disciples – those who do as he commands – not just to make converts.

Jesus' view of the kingdom of God included God's people in God's place: land is important. This included "the renewal of all things" (Matthew 19:28).

For Jesus and other Jews, salvation was not about going to heaven when you die but rather the establishment of God's rule.

Loving people means loving the world in which they live; the two are not at odds with each other.

Jesus reconciles all things to himself, including non-human creation.

Climate change is a moral failure, one that threatens human beings and the world that we all share. However, what does this have to do with the Church? Some Christians protest that Christianity is all about making converts. Didn't Jesus give us a Great Commission to save souls in Matthew 28? Don't we detract from the job of doing this if we get tied up in issues like climate change, even if it really is happening? By now it should be clear that this can't entirely be the case, since our original mission was to represent God to the cosmos and to undertake the wise rule of the Earth. Since we are being remade in the image of Christ, we should be doing all the sorts of things we were originally called to do.

In Matthew 28:18-20, Jesus says:

> All authority in heaven and on Earth has been given to me. Go therefore and make disciples of all nations, baptizing them in the name of the Father and of the Son and of the Holy Spirit, and teaching them to obey everything that I have commanded you. And remember, I am with you always, to the end of the age.

It is clear that Jesus has all authority in heaven and on Earth. The nature of his rule then is not simply about what people identify as 'spiritual' issues like personal morality, prayer and evangelism. Jesus' authority is over such things as politics, sport and the environment.

Furthermore, Jesus calls us to make disciples, not just converts. All that stuff Jesus said about life in the kingdom matters a lot, because we are to teach people to obey all of it. That stuff about not having anything to do with the economic system of empires such as Caesars and giving the whole of ourselves to God (Mark 12:13-17). That stuff about non-violent resistance against illegitimate power like that of Rome (Matthew 5:38-42). That stuff about how to be God's new people and a city on a hill, an alternate society (Matthew 5:14).

We need to understand where Jesus saw himself with regards to God's unfolding story of putting the world right, and how that included "the renewal of all things" (Matthew 19:28), which includes the non-human creation. When Jesus proclaimed the Gospel, what did he mean? In Isaiah 40, the announcement of the Gospel includes the proclaiming of the forgiveness of Israel's sins (verses 1, 9), the coming of God (verses 3-5), and the gathering in of his flock (verses 10-11).

So what did Jesus mean when he proclaimed the Gospel and called for repentance - literally a change of mind? It doesn't mean simply a call to a personal decision for Christ, though it can't mean anything less than that. It also means that the kingdom of God was not fully manifest: Israel had returned, but the people still did not have the law of God written on their hearts as Ezekiel had promised

(Ezekiel 36:26-28). In other words, sin still ruled through individuals and institutions and had to be overcome! Only once sin was dealt with could the people be gathered.

This doesn't mean that land and earthly existence is not the end goal. Far from it! The very evidence that Jesus had overcome sin, evil and death, was that he was physically raised from the dead. The interest of the Jews was in land – physical and solid – and in lives richly and well lived. This should make it clear to us that people are not saved from the Earth but should instead expect to be renewed with the Earth: God's people in God's place. A well thought-out resurrection theology should make it clear that being in God's place is not being out of the body in heaven, but rather having a resurrected body upon a renewed Earth.

Loving people by loving creation

If God truly loves the world, then we are to love it as well. This means people, and the world in which they live. People suffer because of environmental impacts such as climate change.

It is sometimes suggested that caring for the environment comes at a cost of caring for people. A Rocha is a Christian conservation organisation established in over 19 countries.[18] A Rocha's projects are deeply contextual and aimed at healing people as well as habitats. In Kenya, forests are protected and mangroves planted, and communities educated to protect their natural resources. Eco-tourism is developed and money used to provide bursaries for local students to pursue a secondary school education. Unlike some unhelpful caricatures of environmental work, in this case the environment is not put before people, but the two are wedded together. James reminds us that without caring for people's physical needs, our message of peace is empty (James 2:14-17):

> What use is it, my brethren, if someone says he has faith but he has no works? Can that faith save him? If a brother or sister is without clothing

---

18 See their website www.arocha.org. You can also read about their history in Peter Harris, *Kingfisher's Fire: A story of hope for God's Earth* (Oxford: Monarch Books, 2009).

and in need of daily food, and one of you says to them, "Go in peace, be warmed and be filled," and yet you do not give them what is necessary for their body, what use is that? Even so faith, if it has no works, is dead, being by itself.

How can we go about saving souls while our very lifestyles threaten people's physical well being? Will they listen to us? Will our words be full of meaning and love for our hearers, like a beautiful song – or will it sound hypocritical and shallow, like the sound of clashing cymbals? The Bible calls us to holistic mission: to speak and live the Gospel, practising peace, justice and creation care.

It is sometimes overlooked that God is committed to the whole of creation and is seeking to restore it in its entirety under Christ (see also Ephesians 1, Colossians 1). These passages are sometimes said to refer to the 'Cosmic Christ' because of the all-encompassing nature of his redeeming work.

Let's follow the ideas presented in Colossians 1:15-23. *All things* were created through Jesus and for him (verses 15-16). *All things* are maintained in Jesus (verse 17). He is the firstborn from the dead and the head of the church (verse 18). In him, the fullness of God dwells – to see Jesus is to see God (verse 19). The punch-line is that just as *all things* were created by God through Jesus, so *all things* are reconciled to God through Jesus (verse 20), and we cannot see this as being limited to humans only (otherwise the text would have read "all people"). In verses 21-23, this reconciliation is brought down to level of the church at Colossae, but is not limited to them, or us, alone.

> He is the image of the invisible God, the firstborn of all creation; for in him all things in heaven and on earth were created, things visible and invisible, whether thrones or dominions or rulers or powers—all things have been created through him and for him. He himself is before all things, and in him all things hold together. He is the head of the body, the church; he is the beginning, the firstborn from the dead, so that he might come to have first place in everything. For in him all the fullness of God was pleased to dwell, and through him God was pleased to reconcile

to himself all things, whether on earth or in heaven, by making peace through the blood of his cross.

And you who were once estranged and hostile in mind, doing evil deeds, he has now reconciled in his fleshly body through death, so as to present you holy and blameless and irreproachable before him—provided that you continue securely established and steadfast in the faith, without shifting from the hope promised by the Gospel that you heard, which has been proclaimed to every creature under heaven. I, Paul, became a servant of this Gospel. (Colossian 1:15-23)

So in Christ, *all things* will be reconciled. This means that we can't build paradise on Earth through our own efforts. When Christians live lives of peace, justice and restoration of relationships and creation, we provide a foretaste of what is to come. One might say that humans can save the whales (to invoke a conservation cry from a generation ago) but only God can save the Earth. Christians acting to save the whales, or the climate system, are signs that God is yet coming to save the Earth.

1.5 Resisting empire: politics, religion and caring for creation

Key points:

Economic growth is often made into an idol; one which damages creation and humans.

Economic systems, ancient and modern, have been shown to damage creation.

The Gospel opposes empires in all forms, and this can include their economic aspects.

When society idolises money, it distorts the true nature of humanity and expands to overwhelm creation, becoming a self-destructive machine.

The Lordship of Christ addresses the modern myths of western society, including the narratives at the centre of the societal machine.

We have now seen that matter matters to God. Bodies are for resurrecting, creation is for renewing, and land is the place where God's people can be God's people.

Given this, and given that Jesus has authority on Earth as well as in heaven (Matthew 28:18), how can we start to think biblically about the root causes of climate change?

One of the unchallenged assumptions of today's society is that of economic growth. In a neo-liberal worldview, which is the dominant political view in the West, the economy is placed above the welfare of individuals, ultimately of society as a whole, and the environment. This is not to say that there is anything inherently wrong with trade, buying and selling, making money and so on. It is because at some point the economy has ceased to be something that operates for people but instead becomes something that enslaves them. This happens when we effectively worship the economy and capitalism becomes what Brian McClaren calls "theocapitalism".[19] Claire will tease this out more in Section Three. For now, we will consider a theology of empire.

The Gospel is clearly an anti-empire message by default: no Lord but Jesus! This is well captured in Jesus' saying about money that "You cannot serve both God and Mammon" (Matthew 6:24). While empire has come to be a loaded and contested term, a useful definition is:

> a political and economic order that succeeds in subsuming previously disparate nations and economies under a rule that can call on both the "hard power" of military might and technological achievement and the "soft power" embedded in deep structures of ideology, philosophy and theology.[20]

The anti-empire nature of the Gospel needs a little unpacking, but it is important to understand it. When authors of the New Testament discussed "the Gospel" (*euangelion*), they were using a word that had biblical as well as contemporary currency. In Isaiah 40, "good news" is proclamation of the forgiveness of sins, the coming of God and freedom from foreign rule. N.T. Wright contends that even though the people had returned to the land, they were never free of foreign rule. Hence, the Gospel message is anti-coloniser and anti-empire. Read the following passage about Augustus Caesar and ask yourself how a Christian reader could ever understand the Gospel as purely spiritual and apolitical?

---

19 Brian McClaren, *Everything must change: Jesus, global crisis, and a revolution of hope* (Nashville: Thomas Nelson, 2007), 190.
20 Andy Crouch in Scott McKnight and Joseph B. Modica (eds), *Jesus is Lord and Caesar is Not: Evaluating Empire in New Testament Studies* (Downers Grove: IVP Academic, 2013), 8.

a saviour for us and those who come after us, to make war to cease, to create order everywhere ...; the birthday of the god [Augustus] was the beginning for the world of the *glad tidings* that have come to men through him ...(emphasis added).[21]

There are a number of key words and ideas in this quote that are echoed in the New Testament. Caesar is a saviour, one who makes peace (wars to cease), whose birth is to be celebrated, who is good news (glad tidings). Hence the start of Romans, as with the start of Mark, is a declaration that an alternative Lord and Saviour is on offer in the form of a Jewish carpenter who claimed to be Israel's king and acted just as if he were Israel's God.

> The Gospel concerning his Son, who was descended from David according to the flesh and was declared to be Son of God with power according to the spirit of holiness by resurrection from the dead, Jesus Christ our Lord (Romans 1:3-4).

> The beginning of the good news of Jesus Christ, the Son of God (Mark 1:1).

Now, politicians rarely claim such exalted status as is claimed for Augustus (at least in the West), but there are grand claims made to solve society's ills through economic management, austerity or aggressive foreign policy.

That aside, I want to suggest the treatment of our economic system as our 'saviour' in exactly this sort of way is part of the cause of climate change. Let's consider Rome again. The Roman political machine has been summarised in two words: war and law. Many Romans knew the empire made large profits out of warfare and expansion. Rome lived in luxury while the rest of the empire lived in relative or actual poverty. In his article *Taxes and Trade in the Roman Empire (200 B.C.-A.D. 400)*, Keith Hopkins observes that

> the model implies an increased monetization of the Roman economy, the commercialization of exchange, an elongation of the links between producers and consumers, the growth of specialist intermediaries (traders, shippers, bankers), and an unprecedented level of urbanization.

---

21 Tom Wright, *What St Paul Really Said* (Oxford: Lion Books, 2003), 43.

The model illustrates the close connection between changes on the level of individual action by simple peasants and relatively large-scale changes, such as the growth of towns.[22]

There are obvious parallels with modern capitalism.

Rome's economic and military activity also had an impact on the environment. Rome was responsible for significant deforestation due to timber harvesting for construction and metal smelting. A major source of timber use was charcoal making for use with kilns. Hughes and Thirgood in their paper *Deforestation, Erosion, and Forest Management in Ancient Greece and Rome* also note that "Forests supplied wood not only for ships but also for chariots, battering rams, and other huge siege engines, and stock for a host of weapons."[23]

Through a complex chain of physical causation, this deforestation led to an increase in malarial infections, as well as flooding, river mouth silting and soil erosion in the vicinity of Rome. Erosion was widespread in ancient Rome and Greece, as well as microclimate change, leading to a decline in agricultural production[24].

In the modern world, global capital functions as empire. Divinity professor Harvey Cox, in an article in *The Atlantic Online*, identifies the market as "God". He sees strong parallels between "descriptions of market reforms, monetary policy, and the convolutions of the Dow and the Romans." [25] The views put forward by economists are "pieces of a grand narrative about the inner meaning of human history, why things had gone wrong, and how to put them right" which Cox identifies as myths of origin.[26]

---

22 Keith Hopkins, "Taxes and Trade in the Roman Empire (200 B.C.-A.D. 400)", *The Journal of Roman Studies*, 70 (November 1980), 101-125.
23 J. Donald Hughes and J. V. Thirgood, "Deforestation, Erosion, and Forest Management in Ancient Greece and Rome", *Journal of Forest History*, 26 (April 1982), 60-75.
24 See for example Joseph Tainter, *The Collapse of Complex Societies* (Cambridge University: Cambridge University Press, 1990), chapter 5.
25 Harvey Cox, "The market as God: living in the new dispensation", *The Atlantic Online*, (March 1 1999), http://www.theatlantic.com/magazine/archive/1999/03/the-market-as-god/306397/, accessed 6 February 2014.
26 Harvey Cox, "The market as God: living in the new dispensation."

The market is said to be mysterious in the same way that aspects of the character of God are. Ultimately, the market knows best. Cox sees this elevation of the market to the status of "God" as happening at the same time as the decline of the influence of other institutions, such as the church, that once kept it in its place. Now, everything can be bought but nothing has inherent worth, including human beings. Humans are merely part of the market and are little regarded by it. For example, the market can be gleeful about the expansion of cigarette sales to children in Asia,[27] and thrive on the manufacture of weapons designed to kill people.

Finally, it appears to Cox as if nothing is sacred or unmarketable, suggesting that

> such previously unmarketable states of grace as serenity and tranquillity are now appearing in the catalogues. Your personal vision quest can take place in unspoiled wildernesses that are pictured as virtually unreachable – except, presumably, by the other people who read the same catalogue. Furthermore, ecstasy and spirituality are now offered in a convenient generic form. Thus the Market makes available the religious benefits that once required prayer and fasting ....[28]

Brian Walsh and Sylvia Keesmaat echo much of this sentiment in *Colossians Remixed*, linking the imagery of Caesar on coins, standards, artwork etc to that of modern day advertising.[29] They highlight the ubiquity of the corporate logo and advertising in all aspects of our culture, from billboards to clothing, public transport and toothbrushes. Some, like the Disney Corporation, are extremely good at placing their logo on everything conceivable. Other marketing ploys are far from subtle: Coke adds life, and four wheel drives are not functional vehicles but rather enable you to escape from the fast-paced nature of city life into the wide world... even if you never do! Behind the imagery are sweatshops, inequality, violence and environmental degradation. Often the imperial goals are far less subtle. The CEO of Walmart once said that "Our priorities are that we

---

27   Harvey Cox, "The market as God: living in the new dispensation."
28   Harvey Cox, "The market as God: living in the new dispensation."
29   Brian J. Walsh and Sylvia Keesmaat, *Colossians Remixed: Subverting the Empire*, (Nottingham: IVP Academic, 2004), 63.

want to dominate North America first, then South America, and then Asia and then Europe."[30] It is hard to be any more blatantly imperial than that!

Brian McLaren in *Everything Must Change* identifies four major cogs of society (which he calls the societal machine). The first three are the prosperity, security and equity systems.[31]

The prosperity system produces all of the goods designed to make us happy, those things that make us thrive rather than simply survive. The various subsystems of the prosperity system include agriculture, manufacturing, energy, entertainment, and communications. McLaren warns that because some have a bigger share of these products or services, jealousies erupt and so in order to protect their share of prosperity, governments and individuals will desire security in order to protect their pursuit of happiness. The security system consists of subsystems like intelligence, border control, policing, and weaponry. They also require recruitment, training and infrastructure support and can be very expensive to maintain. This need for financing and recruitment means that an equity system is required to spread the costs of the security system and spread the benefits of the prosperity system.

McLaren reminds us that the societal machine exists *inside the Earth's ecosystem*. When the societal machine expands it overwhelms the ecosystem (resource depletion, global warming) and begins to show signs of breakdown (famine, retributive violence, war). Thus it becomes the "suicide machine". That is, the system begins to self-destruct. McLaren maintains that this has already begun to happen.

The fourth cog in the societal machine is the "framing story" or "grand narrative". The story for the prosperity system is dysfunctional because it is focussed on economic growth without regards to natural limits. The Christian counter-narrative is the Gospel. It addresses the various false stories based on human hubris, aggression, greed or insignificance. Just as the Gospel was a counter-

30 Walsh and Keesmaat, 166.
31 Brian McLaren, 54.

empire message in the first century, so it is today. The Gospel in its fullness proclaims a message of a God who is faithful to the promises he has made to his creation, to restore relationships between himself and humanity, humanity within itself, and humanity and the rest of creation. In dealing with sin and evil on the cross, God reconciles humanity to himself in order that humanity might finally enter God's rest, his shalom, as may the whole of creation.

From this perspective, it is not surprising that evangelical elder John Stott should have included a chapter on creation care in his book *The Radical Disciple*[32] and that the activist theologian Ched Myers emphasises the importance of "Sabbath Economics".[33] The creation continues to operate without human labour, and God will continue to provide for the good life. The Sabbath rest contrasts strongly with the dysfunctional prosperity system where, in the words of Citigroup, "business never sleeps".

Sabbath rest was for all people to rest from their labours, but also the land. Every seventh year, the land was to have its own Sabbath rest to allow the soil to recover from over-use (Exodus 23:9-11). Likewise, the Sabbath law included rights for wild animals so that they might be able to fulfil their own creation mandate (Genesis 1:22 compare Exodus 23:11). Psalm 104 urges us to keep some wild places undeveloped because they matter to God. Science also tells us this is valuable so that they may continue to provide the priceless services to humanity that climate change so threatens. For example, trees store climate-warming carbon and their roots purify water in catchment areas. Coral reefs are not only tourist attractions, they provide valuable sources of protein for Pacific Island communities, as well as hosting hundreds of species which are an important part of creation in their own right.

Christians of all people should not buy into stories of economic growth and freedom and associate them with spiritual freedom. Wealth may very well be 'earned' by our hard work, but all good things come from God (James 1:17)

---

32 John Stott, *The Radical Disciple* (Nottingham: IVP, 2010), chapter 4.

33 See for example Ched Myers, *The Biblical vision of Sabbath economics* (Washington DC: Tell the Word, 2002). See also his website www.chedmyers.org.

and are not simply for us to hoard. Exactly how economic growth, freedom and justice (e.g. fair redistribution of wealth) are achieved is often argued along political ideologies which are labeled as being either 'left' and 'right'. It is time for many of these concepts to be set aside within the Church. Economic growth does not equate with freedom without remainder. Outdated concepts of left and right hinder us from discovering biblical principles of maintaining a sustainable and just society.

In Galatians 5:13-15, freedom is not for self-indulgence. Rather it is for love, which sums up our whole duty to each other. The opposite of this is selfish ambition and partisan infighting – the world sees us squabble over how to (or sadly in some cases whether or not to) share our wealth – to the detriment of the Gospel.

> For you were called to freedom, brothers and sisters; only do not use your freedom as an opportunity for self-indulgence, but through love become slaves to one another. For the whole law is summed up in a single commandment, "You shall love your neighbor as yourself." If, however, you bite and devour one another, take care that you are not consumed by one another. (Galatians 5:13-15)

The time has come for the Church to realise that the Lordship of Christ addresses the modern myths of Western society, and that as the new body politic the way to fulfil the call to be a blessing to the world (Genesis 12:1-3) is to challenge Caesar wherever he may be found. It is therefore impossible to grasp the true nature of the challenge of climate change without first understanding its political nature, namely that that the abuse of power, in particular by empires of various sorts (governments, businesses, etc.) is to blame. That power of empire is to be resisted by the church. Jesus presents us with another way of exercising power that challenges all systems that put profit and the raw exercise of power above God, people and creation: "For the Son of Man came not to be served but to serve, and to give his life a ransom for many." (Mark 10:45)

1.6 More than Good Samaritans: the ethics of climate change

Key points:

The parable of the Good Samaritan shows us we have to care for those who suffer the impacts of climate change, because in a globalised economy with one atmosphere, **everyone** is our neighbour.

People are suffering from climate change right now, especially those in developing nations.

Just as a ruthless economic and political system produced the bandits in the parable of the Good Samaritan, a broken economic and political system is producing climate change. We need to challenge and change this system.

Part of challenging the system is challenging its messages, especially the public relations of those with vested interests.

Just then a lawyer stood up to test Jesus. "Teacher," he said, "what must I do to inherit eternal life?" He said to him, "What is written in the law? What do you read there?" He answered, "You shall love the Lord your God with all your heart, and with all your soul, and with all your strength, and with all your mind; and your neighbour as yourself." And he said to him, "You have given the right answer; do this, and you will live."

But wanting to justify himself, he asked Jesus, "And who is my neighbour?"

Jesus replied, "A man was going down from Jerusalem to Jericho, and fell into the hands of robbers, who stripped him, beat him, and went away, leaving him half dead. Now by chance a priest was going down that road; and when he saw him, he passed by on the other side. So likewise a Levite, when he came to the place and saw him, passed by on the other side. But a Samaritan while travelling came near him; and when he saw him, he was moved with pity. He went to him and bandaged his wounds, having poured oil and wine on them. Then he put him on his own animal, brought him to an inn, and took care of him. The next day he took out two denarii, gave them to the innkeeper, and said, 'Take care of him; and when I come back, I will repay you whatever more you spend.' Which of these three, do you think, was a neighbour to the man who fell into the hands of the robbers?" He said, "The one who showed him mercy." Jesus said to him, "Go and do likewise." (Luke 10:25-37)

It is essential that we consider the implications of Jesus' parable of the Good Samaritan within the context of empire. The parable comes after an expert in the Law of Moses tries to test Jesus and limit the understanding of who his neighbour was (i.e. who he was responsible for). What he received in response was a story about what kind of neighbour he was to be, and a suggestion that his neighbour might even be his enemy.

So who is our neighbour? We need to recognise that we live within one atmosphere and a globalised economy. Greenhouse gases know no national boundaries, and the emissions that we produce when we drive a car or switch on a light affect everyone on the globe.[34] More than this, some people are more vulnerable to climate change due to their economy, geography or climate. Often these are people living in the developing world on marginal lands with poor infrastructure and limited ability to adapt to the consequences of climate change. The people of Tuvalu are an obvious example, threatened by sea level rise.[35] Other examples in the developed world demonstrate that we ourselves are not immune, such as the residents of New Orleans during and after Hurricane

---

34 This statement presumes that both of these actions involve the burning of fossil fuels, such as when using a petrol-powered car and electricity that is generated by burning coal. Obviously electric vehicles and solar power provide ways to drive and use lighting that have much lower environmental impact, however wide scale transitions in these directions are only in their infancy.

35 Friends of the Earth International, *Climate Change: Voices from Communities Affected by Climate Change* (Amsterdam: Friends of the Earth, 2007), 32.

Katrina in 2005, or the elevated death rate amongst the elderly in France and Switzerland during the 2003 European heatwave (see more in Section Two).

Our globalised economy means that we can outsource some of our environmental impacts and consequent suffering to countries with more relaxed environmental laws, just as we rely upon cheap labour and manufactured goods from the developing world. In doing so, we simultaneously outsource much of our greenhouse gas emissions overseas. Australia for example, exports coal and raw materials to China and then buys manufactured goods from them. None of this is to say that trade is wrong or to point the finger at developing economies who want to share in our affluence. However, it should be clear that in a warming world, everyone is our neighbour and we need to be a neighbour to everyone. Additionally, given that carbon dioxide lasts in the atmosphere for an incredibly long time, our impact is extended well into the future, making future generations our neighbours as well!

This latter point is worth thinking about a little more. If we think about the rate of global population increase, there are now more people alive today than have ever lived. If population continues to rise exponentially, that represents a lot of people as yet unborn who have no say in what we do, but will be born, grow up, and die (quite possibly prematurely) due to our actions. A world of 4 or more degrees Celsius warmer than at the start of the Industrial Revolution will cause melting of glaciers and ice sheets for thousands of years, and provide a climate that is warmer than pre-industrial levels for hundreds of thousands of years.[36]

A typical approach to the parable of the Good Samaritan would then be to say that we need to bind up the wounds of those are harmed by the impacts of climate change. We resolve to invest in adaptation technologies for the developing world such as sea walls and basic mitigation technologies such as solar-powered stoves. It also means acts of compassion such as accepting environmental refugees, and providing medical aid and disaster relief. With malaria spreading to new parts

---

36 See David Archer, *The Long Thaw: How Humans are Changing the next 100,000 Years of Earth's Climate* (Princeton: Princeton University Press, 2009) and Curt Stager, *Deep Future: The Next 100,000 years of Life on Earth* (Melbourne: Scribe, 2011).

of the world as the planet warms, more mosquito nets and malarial drugs will also be required.

And yet if climate change may result in many deaths as people are hit by droughts, declining crop yields[37] and other climate related disasters, isn't binding up the wounds very far short of what we could or should do? Christians were known for looking after plague victims in the Roman Empire when others would not.[38] They were present in numbers as Roman society faced collapse through malaria and the oncoming barbarians. These Christians faced enormous problems with acts of sacrificial love and service, without necessarily thinking they could "change the world".

This moral imperative to be Good Samaritans applies right now because people are suffering and dying today as a result of a changing climate. Although the science is discussed in Section Two, it is worth pointing out at this point that people are already being affected by sea level rise. The nation of Tuvalu is slowly losing land as a result of it. People are already abandoning the Carteret Islands, off the coast of Papua New Guinea, also due to sea level rise. None of the people under threat want to abandon their homes, and with the right support they may be able to live there for many decades. But is it acceptable to keep placing them under threat by the way in which we live? In other parts of the world, such as the Kenyan highlands, warming temperatures have allowed malaria to spread for the first time.

So yes, there is much binding up of wounds to do right now. But there is so much more that we can, and must, do. Although it is not the point of the parable per se, have you ever wondered why there were bandits in the first place? What was it about the parable that meant that all of the hearers could nod their head in recognition and understanding? Rome taxed her colonies very heavily, and it wasn't difficult for landowners to find themselves in debt. Many ended up

---

37 Some crops have been declining in yields at a rate of a maximum of 5% per decade. See Christopher B. Field et al., *Climate Change 2014: Impacts, Adaptation, and Vulnerability, Summary for Policy Makers* (Geneva: IPCC, 2014). http://report.mitigation2014.org/spm/ipcc_wg3_ar5_summary-for-policymakers_approved.pdf, accessed 26 April 2014.

38 Rodney Stark, *The Rise of Christianity: How the Obscure, Marginal Jesus Movement Became the Dominant Religious Force in the Western World in a Few Centuries* (San Francisco: Harper, 1997), chapter 4.

having to sell their land, and the Herodians and temple elite were only too keen to buy. For people without land or a trade, all they had left was either to become day labourers (as featured in many of Jesus' parables) or beggars (particularly those too sick or infirm to work) or… bandits.

Hence, while human sinfulness means such behaviour cannot be excused, an oppressive regime was the cause of the conditions that led to banditry. This brings us back to the issue of the global capitalist "suicide machine" and issues of labour, poverty, inequality and greenhouse gas emissions. If what I have argued for might seem a bit of a stretch, I am really just teasing out the historical context behind the parable. Jesus was a lot more direct in Matthew 23:23, echoing Micah 6:8 – justice and mercy and faith are weightier than tithing, or any of the other theological gnats we construct.

> "Woe to you, scribes and Pharisees, hypocrites! For you tithe mint, dill, and cummin, and have neglected the weightier matters of the law: justice and mercy and faith. It is these you ought to have practiced without neglecting the others. (Matthew 23:23)
>
> He has told you, O mortal, what is good; and what does the LORD require of you but to do justice, and to love kindness, and to walk humbly with your God? (Micah 6:8)

As the Church, we need to speak truth to power prophetically, just as the prophets of Israel did to their kings and the Apostles did to the leaders of Jerusalem and Rome. This is not just speaking the truth of the Gospel for the purpose of personal conversion (but it is at least that). It is *public ethics*, not just the "pubic ethics" of sexual and reproductive issues that preoccupy many Christians. It isn't just bedroom ethics, but boardroom ethics as well!

Approaching climate change, we advocate not only for more efficient machinery and green energy, but for a more just world. And we do this knowing that a time will come when all things will be put to rights, and all people will be brought to justice. The path of shalom means that while Christians have a voice in the West, we are to used that power for the welfare and flourishing of others. Christian aid work and advocacy are a powerful way in which the rule of Christ is proclaimed, and should be directed at subverting the very system that brings potential disaster.

This truth-telling will mean taking on those who drive global warming. This includes, in particular, the fossil fuel industry. Again, we need to see the link between the commands of personal piety and corporate sin. We might be happy to tell our children not to steal or lie, but what of those who steal from the poor and future generations, and lie about it not being their own fault? The central theme of books such as *Merchants of Doubt* is that a minority of special interest groups have borne false witness to the effects of cigarettes, acid rain, the ozone layer and global warming for the sake of maintaining corporate profit or advancing an ideological agenda.[39]

Global warming is certainly "an inconvenient truth" (to use a phrase made famous by Davis Guggenheim's 2006 documentary by the same name). It appears that greed often takes priority over corporate responsibility. Christians are supposed to discover the truth and tell it, to uncover lies and expose them. Perhaps some of the lies are self-deception, but it is then our job to expose those lies as well. If we do not expose the lies and proclaim the truth then we become part of a lie. And given the potentially horrendous consequences of climate change, it is a horrendous lie indeed.

---

39 Naomi Oreskes and Erik M. Conway, *Merchants of Doubt* (New York: Bloomsbury Press, 2010).

1.7 It's the end of the world as we know it (and I feel fine): Does the world have a future?

Key points:

2 Peter 3 doesn't describe the literal burning up of the Earth but its refining. In the same way the Flood was an act of judgment and salvation, "uncreation" and re-creation, the fire in 2 Peter 3 talks about salvation as well as judgment.

There is no such thing as the Rapture. 1 Thessalonians describes believers welcoming Jesus back to the Earth to rule over it.

Heaven is not an escape from Earth. Instead, heaven means the place where God rules – and God's rule will one day be fully revealed on this Earth.

One final piece of the theological puzzle is needed in demonstrating that Christians should be concerned about climate change, for some might ask, "Doesn't the Bible say that that the world is going to be destroyed? Doesn't that fact render climate change a moot point?" Well-known pastor Mark Driscoll said during the Catalyst conference in 2013, that "I know who made the environment and he's coming back to burn it all up. So yes, I drive an SUV."

> First of all you must understand this, that in the last days scoffers will come, scoffing and indulging their own lusts and saying, "Where is the promise of his coming? For ever since our ancestors died, all things continue as they were from the beginning of creation!" They deliberately ignore this fact, that by the word of God heavens existed long ago and an

earth was formed out of water and by means of water, through which the world of that time was deluged with water and perished. But by the same word the present heavens and earth have been reserved for fire, being kept until the day of judgment and destruction of the godless.

But do not ignore this one fact, beloved, that with the Lord one day is like a thousand years, and a thousand years are like one day. The Lord is not slow about his promise, as some think of slowness, but is patient with you, not wanting any to perish, but all to come to repentance. But the day of the Lord will come like a thief, and then the heavens will pass away with a loud noise, and the elements will be dissolved with fire, and the earth and everything that is done on it will be disclosed.

Since all these things are to be dissolved in this way, what sort of persons ought you to be in leading lives of holiness and godliness, waiting for and hastening the coming of the day of God, because of which the heavens will be set ablaze and dissolved, and the elements will melt with fire? But, in accordance with his promise, we wait for new heavens and a new earth, where righteousness is at home. (2 Peter 3:3-13)

Although Driscoll later retracted his statement, claiming it was a joke, his comment was shared across the internet with approval by many Christians. Regardless of Driscoll's real understanding of 2 Peter 3:3-13, this incident illustrates that many hold on to less-than-biblical doctrines of creation and the End Times. They may believe it's the end of the world as they know it, but they feel fine (as the REM song goes) rather than unsettled, which would be a much more appropriate response!

Let's assume for a moment that the Bible tells us that God is coming to scrap the whole creation project and start from scratch, or even that physical creation is to be abandoned for heaven with harps, wings, fluffy clouds etc. Would creation's non-future justify the sentiment behind the Driscoll quote? Hopefully I have already established above that this would not justify the mistreatment of creation. Not only would it be the result of sinful patterns like greed, the worship of money, status and power, it would be an abdication of our divine status as

God's image-bearers in the world, and make us into bandits rather than Good Samaritans.

But does the Bible really teach that the creation itself has no future? Is there truth in Driscoll's throwaway one-liner about 2 Peter 3? There are three ways of thinking about this.

The first thing to note is that Jewish literature contained a genre known as apocalyptic literature. Apocalyptic literature uses highly symbolic language to describe this-worldly events, or as Tom Wright puts it "to evoke the cosmic or theological *meaning* of events in the space-time world by means of a sometimes complex system of metaphors".[40] For example, when the prophet Isaiah (13:10) wrote of the destruction of the city of Babylon, he said that "the sun will be dark at its rising, and the moon will not shed its light".[41] Likewise, when Jesus spoke of the Jerusalem temple being destroyed in Mark 13, he described it being accompanied by earthquakes. Neither comment was meant to be understood as astronomy or geology, simply that both events were metaphorically "earth shattering". So an expectation that the fire in 2 Peter 3 should necessarily be a literal fire is misplaced.

The second thing to note about this passage is the comparison between the return of Christ and the fire, with the Flood of Noah. As we discussed earlier, in the Flood, the waters above the earth and the oceans which were separated at creation come together in an act of "uncreation". The world returns to formlessness. This is not a material destruction as we would think of it, a complete obliteration of the Earth. Instead, the proper functioning of the creation is disrupted. Furthermore, Peter fails to mention the Ark, by which both humans and non-humans were saved from destruction. He also doesn't mention that the waters eventually retreated in an act of re-creation. He failed to mention these things because they were understood by his readers. Does this drawing of parallels between "destruction by fire" and the Flood help us in understanding

---

40 Tom Wright, *The Millenium Myth* (London: SPCK UK, 1999), 27.
41 Tom Wright, *The Millenium Myth*, 27.

Peter's approach to the "end of the world"? Certainly it does. It means that the destruction by fire is not an obliteration of what is, but an apocalyptic way of talking about its purification and renewal.

This understanding of purification and renewal is made clear when we think carefully about how we are to understand verse 10 and the phrase "the elements will be dissolved with fire, and the earth and everything that is done on it will be disclosed." The word 'disclosed' reflects, as Bouma-Prediger notes, the Greek word *heurisko* from where we get our English word heuristic, which is a method of finding things out.[42] This helps shape our understanding of the role of the fire and what it means for the elements to be dissolved. In this reading, the fire is for purification of the Earth from sin, and not for the physical Earth's actual destruction. This idea of a process of refining is alluded to back in 1 Peter 1:7 where Peter writes "so that the genuineness of your faith—being more precious than gold that, though perishable, is tested by fire—may be found to result in praise and glory and honor when Jesus Christ is revealed". The phrase "may be found" reflects our word *heurisko* again, and the revelation of Jesus refers to his return to earth as described in the book of 2 Peter.

It is in this context that the language of destruction in verses 11-12 should be properly understood. The day of the Lord's appearing is about the judgment of deeds: the works are laid bare or revealed (verse 10) so that the new heavens and Earth may only consist of righteousness (verse 13). This has implications for the current lifestyle of the believer – which should be characterised not by the abandonment of the world to its fiery fate, but rather by holy living. The thesis of this book is that this holy living should include an attitude of care toward the Earth, and a call to combat climate change.

The view that the Earth will be physically destroyed usually goes hand in hand with the idea that when Jesus appears again, he will take Christians away before the end comes. We are to be beamed up just like in those Star Trek episodes! In

---

42 Steven Bouma-Prediger, *For the beauty of the Earth: A Christian vision for creation care* 2nd edition (Grand Rapids: Baker Academic, 2010), 68-69.

1 Thessalonians 4 Paul describes what Christ's return will be like. Jesus returns with those who have died in the faith (verses 14-16) and those alive will rise into the air to meet him (verse 17). Rapture theology assumes that we then go to live with Jesus in heaven, even though the text says no such thing. As New Testament scholar N.T. Wright notes in his *For Everyone* commentary, it is important to understand the common secular use of the Greek word translated as 'appearance'. When Caesar appeared at Rome after some great military victory, the dignitaries would go out to greet him – not to then go off somewhere else but rather to welcome him into the city! We are not raptured off to heaven: instead we will join Christ in a new heavens and Earth. The idea of being caught up in the clouds (1 Thessalonians 4:17) is to be expected, as Jesus disappeared behind the clouds when he departed and would return in exactly the same way (Acts 1:9-11). Nothing in the passage and everything about the Greek word for appearance suggest that we leave Earth for heaven.

> Then we who are alive, who are left, will be caught up in the clouds together with them to meet the Lord in the air; and so we will be with the Lord forever. (1 Thessalonians 4:17)

> When he had said this, as they were watching, he was lifted up, and a cloud took him out of their sight. While he was going and they were gazing up toward heaven, suddenly two men in white robes stood by them. They said, "Men of Galilee, why do you stand looking up toward heaven? This Jesus, who has been taken up from you into heaven, will come in the same way as you saw him go into heaven." (Acts 1:9-11)

This idea is also clear in Revelation 20-21, where we read of the city of God descending to the Earth. There is no temple because the city itself is a temple and God is everywhere. Heaven and Earth meet where God's kingdom has come and his will is done.[43] In this passage, the new heavens and new earth replace the old (Revelation 21:1), but this replacement is one of character and not physicality as such – in this new order there is no more death or mourning or pain, for God dwells with people.

---

43 Tom Wright, *Paul for Everyone: Galatians and Thessalonians* (Louisville: Westminster John Knox Press, 2004), 122-126.

> Then I saw a new heaven and a new earth; for the first heaven and the first earth had passed away, and the sea was no more. And I saw the holy city, the new Jerusalem, coming down out of heaven from God, prepared as a bride adorned for her husband. And I heard a loud voice from the throne saying,
>
> "See, the home of God is among mortals. He will dwell with them; they will be his peoples, and God himself will be with them;he will wipe every tear from their eyes. Death will be no more; mourning and crying and pain will be no more,for the first things have passed away."
>
> And the one who was seated on the throne said, "See, I am making all things new." (Revelation 21:1-5a)

A final text that is sometimes used to promote an external future in heaven is John 14:2 where Jesus says "In my Father's house there are many dwelling-places." Jesus says he is preparing a place for his disciples. The Greek word which is translated as *dwelling-places* refers to a temporary resting place or way station for a traveller.[44]

Far from depicting heaven as a place where we will be *for all time*, heaven comes to Earth when Jesus returns. The implication of all of this is that heaven is not an escape from all the difficulties of life, or merely a here-and-now responsibility to live peaceably, justly and caring for creation. Instead, heaven means the place where God rules. And God's rule will one day be fully revealed on Earth. And because we too will be raised from the dead, no deed of peace, justice or creation care we do here and now is in vain (1 Corinthians 15:58).

Finally, we are in a position to look at Romans 8:19-22.

> For the creation waits with eager longing for the revealing of the children of God; for the creation was subjected to futility, not of its own will but by the will of the one who subjected it, in hope that the creation itself will be set free from its bondage to decay and will obtain the freedom of the glory of the children of God. We know that the whole creation has been groaning in labor pains until now (Romans 8:19-22)

---

44 N.T. Wright, *The Resurrection of the Son of God* (London: SPCK: 2003), 446; Howard A. Snyder with Joel Scandrett, *Salvation means creation healed: The ecology of sin and grace* (Eugene: Cascade Books, 2011), 35.

In this passage, Paul talks about how the creation longs for the future revealing of the sons of God (verse 19) while we groan for our sonship (verse 23). The language of sonship represents the idea of inheritance common in the ancient Near East, where the first born son inherited their father's wealth. Our inheritance is not limited to men, but to all children of God (verse 21). This inheritance is of the new heavens and Earth, just as Jesus said that the meek (those Christ-like in character) would inherit the Earth (Matthew 5:5). We come into this inheritance when we are raised from the dead, also described by Paul as the redemption of our bodies (verse 23).

The creation waits eagerly for us to be raised from the dead, because something about renewed humanity will be beneficial for it. In our renewal, creation will find its own liberation (verse 21). Creation was subjected by God to futility in hope of its eventual liberation, just as humanity was given over to sin, (Romans 1:18-32) ultimately in hope of the resurrection (Romans 8:23-24).

Adam's task was to tend and care for creation and his rebellion and the idolatry of our forebears has meant that creation has not been under the rule for which it was intended. Israel failed to carry out its divinely appointed task, but instead constantly fell into idolatry and dishonoured God's name (Romans 2:17-24).

The Gentile nations also worshipped the creation rather than the Creator. God sent his own Son to fulfil the requirements of the old covenant and be the instrument of the blessings promised to Abraham and his descendants (Genesis 12:1-3). Being in Christ through the Spirit, we now take up this role to bring blessing to the world, proclaiming Jesus as Lord and being peacemakers (Matthew 5:9). When we are finally revealed as sons and daughters of God at the resurrection, then the peace of God will extend to all creatures (Colossians 1; Ephesians 1). This is pretty meaty theology. If it is new to you then you might like to take some time to read through these passages more thoroughly. You might just find that your theological world gets turned upside-down in the most surprising and wonderful way!

Conclusion
Climate Change: a consequence of sin

The claim that human beings are changing the climate due to the burning of fossil fuels is consistent with a biblical doctrine of human sinfulness. We are not at liberty to ignore the science simply because many climate scientists are not Christians. One does not have to be a Christian in order to understand the workings of the world.

Equally, we are not at liberty to ignore the science because such inconvenient truth demands a response. Wilful ignorance is sin. The doctrine of human sinfulness allows for the idea that we can do damage to the planet. Regardless of whether God permits or directly wills damage to the world via climate change, we are not absolved of blame and accountability; the time will come for the destruction of those who destroyed the Earth (Revelation 11:18).

A biblical doctrine of creation informs us that creation matters to God. Creation is a temple for God to dwell in, for us to serve him in, and has a very certain and glorious future at the resurrection. Climate change resulting from human sin and idolatry is an offense to God. It threatens both the non-human creation and our fellow humans who have been made in God's image, making climate change not simply an issue of science or the environment, but also of human rights.

If we believe that God is a just and loving God who calls us to do justice and love the world, then we must speak out and act to mitigate against the effects of climate change. And as we will learn in the next section, the science says we must do so now.

Section Two: The nature of science
Dr Mick Pope

"I love a sunburnt country ... Of droughts and flooding rains."
Dorothea Mackellar, My Country

2.1 Science works!

Key points:

Scientific theories are our best working models, not simply guesses to be treated lightly.

Real science uses scepticism; denial ignores the facts.

The Intergovernmental Panel on Climate Change does a good job of synthesising the present science, but science advances quickly.

Before launching into the details of climate change science, it is necessary to discuss briefly what science is (and what it isn't). This doesn't mean we have to dig too deeply into the history and philosophy of science, but it is certainly worthwhile thinking about it. I touched a little on this in the first section on theology, because for some Christians science can be a touchy subject. But more than that, while science is a method, it often carries with it a philosophy or worldview.

So... what is science?

Science is a method for pursuing understanding about the way the world works. More specifically, it is a set of methods that differs for different fields of study. For example, palaeontology - the study of fossils, uses very different techniques to most fields of physics, simply because while many physics experiments are repeatable, the history of life on Earth is not! Some of the elements of the scientific method include the use of observations, mathematical analysis such as statistics to quantify the significance of measurements, as well as theories. Theory is a much misunderstood word; it's what some Christians use to describe evolution or climate change so as to dismiss them as 'just a theory'.

A theory in science is a working hypothesis that is believed to explain what we observe, and make useful predictions. It is a work in progress; best explanation; open to revision; approximation to the truth. Some scientists might think they are discovering 'the truth' in some Platonic sense, and in the case of mathematics, they might be right. But for most things, scientists know that models are approximations. We look for 'beyond reasonable doubt' rather than absolute certainty. One might suggest that such a way of thinking about knowledge shows more similarities between science and religion than some are willing to admit.

Scepticism is a key tool in real science. I can recall a time while undertaking my PhD when I was convinced that a certain result would follow, yet the more I probed my data, the more wrong I was. I wasn't being sceptical enough about my own theory, but reality provided, well… a reality check! Cherry-picking of results rather than a careful regard for reality is what marks a climate change denier from a climate change sceptic. Yes, it is possible for a scientist to get into a rut, like Albert Einstein's insistence that "God does not play dice" and hence his wasting many years trying to prove that Quantum Mechanics was wrong. Yet science is a communal activity. The emphasis on writing scientific papers ensures that by sharing ideas they might be critiqued and sharpened by a community. The process of peer review has been around since 1967 when it was introduced to the journal *Nature*. While not hole-proof, I can speak from both sides of the fence in that I have both peer-reviewed papers and worked through the process of making corrections and additions in response to the work of reviewers. I can say that it does generally help keep things that are wrong or trivial out of publication, and improve what is eventually published. It can be a slow process, but this isn't always a bad thing.

The reason for going through this sort of preamble is to get us thinking about the role of a body like the Intergovernmental Panel on Climate Change (IPCC). As a summary of the peer-reviewed literature conducted every six years, it is a slow, ponderous but careful process of going over the peer-reviewed literature by experts in the field. It certainly isn't some half-cocked, shoot-from-the-hip statement. It isn't a collection of sound bites, and it isn't funded by special interest groups. While its pronouncements don't come down off Mount Sinai engraved in stone, they do carry a lot of weight (and I don't just mean the size of the print editions). A total of 209 lead authors and 50 review editors from 39 countries, with more than 600 contributing authors from 32 countries contributed to the recently released Fifth Assessment Report (AR5) of the basic science, with another two large volumes on mitigation (how to avoid climate change), and impacts, adaptation and vulnerability.

Given the scope of the IPCC assessment process, two observations follow. Firstly, when the occasional error is found, we should be surprised at *how few* errors there are and not *how many*. Secondly, given the timescale and consensus nature of their reports, we should expect that knowledge is growing at a faster rate than these reports can allow for, and thus the reports are probably a little more conservative than they could be. Therefore, in what follows I will refer to AR5 for the basics, but will look to other reports and papers, older or more recent, when necessary.

Comment by Claire: I'd suggest a third observation here – that there is also a "systematic tendency to understatement" as identified in the *Garnaut Climate Change Review*. In the case of published climate change research, "scholarly reticence" is common: often politically relevant or convenient information is given priority over that which is disconcerting or alarming. In the conclusion of his report, Prof. Ross Garnaut suggests that on this basis it would be prudent to prepare ourselves for even more urgent calls for action in the future.[45]

So, what next? After thinking a bit about the differences between weather and climate (which often confuses people), we will examine three different areas of climate science. This examination can't be comprehensive – the field and evidence is just too vast.

---

45 Ross Garnaut, *Garnaut Climate Change Review – Update 2011, Update Paper five: The science of climate change* (Canberra: Commonwealth of Australia, 2011) 53-54.

Firstly, we will look at how we know the planet is warming, including some of the impacts that are occurring now. It is important to realise that lives are being impacted now and people have already died because of climate change. Secondly, we will look at how it is that we know humans are largely responsible for changing the climate. Finally, we will look into possible futures and consider the challenge of cutting our greenhouse gas emissions. We won't look at solutions – they will be covered in Section Five – but will highlight just how drastic the changes are going to have to be.

In what follows, I am assuming most readers do not have degrees in climate science as I do. But rather than saying, "just trust me, I'm a scientist", what I am asking you to do is what former Supreme Court Judge David Harper had to do: base your judgments on "a rigorous examination of the evidence"[46]. Ultimately it is not a matter of whether the deniers are right or wrong (although I hope to show that the deniers are in fact wrong, and that the basic science is settled enough for us to act)! Rather, it is important to appreciate that the risks to which so many reputable scientists point are so significant in their scope and impact that some presence of doubt in no way justifies complacency or inaction. Indeed, the risks are so great that ignoring them is just not an option.

---

46 David Harper, "Delivering Judgment on the Great Global Warming Debate", *The Age*, February 8, 2014. http://www.theage.com.au/comment/delivering-judgment-on-the-great-global-warming-debate-20140207-32743.html, accessed 20 October 2014.

2.2 Some like it hot

Key points:

Climate is what you expect, weather is what you get.

We forecast weather but make climate projections: they are different statements on different timescales.

Climate "variability" describes natural changes whereas climate "change" refers specifically to the impacts of humans.

Four different sets of climate records show that the planet has warmed since the start of the Industrial Revolution.

The last three decades are the warmest for at least the past 1,400 years.

Changes in solar output and likely changes in the El-Niño Southern Oscillation can explain why global average temperatures have increased more slowly over the past decade while continents like Australia continue to break temperature records.

The oceans absorb more heat than the atmosphere and have continued to warm, with unprecedented levels of warming below 700m.

The present warming goes against the expected longer term cooling trend.

Arctic summer sea ice is at its lowest extent for some 1,400 years.

In a TV documentary produced a number of years ago by Channel 4 in the UK, the father of modern Chaos theory, Edward Lorenz, described the difference between weather and climate as "climate is what you expect, weather is what you get." If I make a prediction about tomorrow's maximum temperature, I am making a weather forecast. However, if I make a statement about the likely global average temperature for 2100, I am making a climate projection. It's all a matter of timescale, and often also detail. To get a little more technical and quote the American Meteorological Society *Glossary of Meteorology*, climate is:

> The slowly varying aspects of the atmosphere–hydrosphere–land surface system. ... It is typically characterized in terms of suitable averages of the climate system over periods of a month or more, taking into consideration the variability in time of these averaged quantities. Climatic classifications include the spatial variation of these time-averaged variables.[47]

With this in mind, we need to note a couple of things. Firstly, weather forecasts are driven very much by a 'as precise as possible' knowledge of the present conditions, and an 'as accurate as possible' representation of how the atmosphere works. This is a different problem to making assessments about climate change, and it is nonsense to jump from anecdotal criticisms about forecast accuracy, to a denial that we simply can't trust model projections of climate change. Apart from the fact that weather forecasts in the short term have improved dramatically, climate projections are interested in the drivers of change (greenhouse gas emissions, volcanoes and changes in the sun's output) and how the Earth system responds.

Secondly, we need to distinguish between climate "variability" and climate "change". Since the Earth formed out of a molten ball of rock some 4.5 billion years ago, we've seen oceans form, volcanoes belch out an atmosphere, life emerge, continents rise, fall and move, the sun brighten and cycle through sunspots, asteroids hit the Earth, the Earth's orbit change periodically, ocean

---

47 American Meteorological Society, *Meteorology Glossary*, http://glossary.ametsoc.org/wiki/Main_Page, accessed 12 December 2013.

currents change, and so on.[48] The climate varies naturally over time, from the regular 11 year sunspot cycle, to changes in the shape of the Earth's orbit over millennia. Some changes are cyclical, others irregular. An example of climate variability from the present is the impact of El Niño in Australia or the North Atlantic Oscillation in Europe and North America, which I will explain shortly.

Climate change is understood as the human-caused changes to climate that occur on top of the natural changes we see, principally through the burning of fossil fuels and clearing of land, as well as things like the use of fertilisers and cement production. The major claim being made by climate scientists is that human beings are causing a discernible change to the climate, and that this is clear when comparing the past 150 years or so to climate records over the past few centuries, indeed millennia. Given this combination of natural and human changes, when looking for trends, scientists need to consider carefully the timescales we use. The UK Met Office glossary suggests that 30 years is a good timescale to consider. With this in mind, when we examine temperature trends below, we need to keep in mind that short-term changes do not make a trend, and choice of starting and end points is very important.

Our changing climate

The difference between climate and weather is one area of confusion. Similarly, confusion exists regarding whether we should be talking about climate change or global warming. The burning of fossil fuels emits greenhouse gases, which result in global warming. But temperature is a measure of climate, so a warming climate is a changing climate. Likewise, changes in temperature lead to changes in wind flows and rainfall patterns, so again the climate changes. In this section, we will look at the evidence that the planet is warming. Throughout this book we will be using the term climate change, but remember this includes global warming. Both terms have been in both academic and public use for decades.

---

48   In a book aimed largely at Christians, I need to say now that I am assuming an old Earth, which is a standard scientific assumption. I don't think this is at odds with what the Bible says, but this is an argument for another time. Even assuming a young Earth we can demonstrate that human beings are responsible for the present changes in climate.

We know the planet is warming from direct measurements of temperature from thermometers. As noted in the *Technical Summary* of AR5, global-scale temperature records go back to the mid-19th century.[49] There have been three centres that have collected and analysed surface temperatures over land and sea: the UK Met Office Hadley Centre together with the Climatic Research Unit in Norwich, NASA's Goddard Institute of Space Studies, and National Oceanic and Atmospheric Administration's National Climatic Data Centre in the USA. When you compare their various datasets, they tell the same story of a warming world since about 1850.

Imagine however that you assumed some kind of conspiracy or, at the very least, incompetency in the climate scientist community. Imagine if you took the deniers' concerns very seriously, started a project run by a physicist from an independent, non-government, non-corporate funded research body, and let statisticians analyse the data. Imagine if you used raw data rather than homogenised or edited data. What might happen when you did this?

Well, in 2010, the Berkeley Earth did just this (see berkeleyearth.org). By 2012 they had shown that previous datasets, which revealed the story of a dramatically warming world, were essentially... wait for the drum-roll... correct!

Now, how many more times does this exercise need to be repeated to show that the planet is warming? It is time to move beyond paranoia and denial and look at what the data actually shows us.

---

[49] T.F Stocker et al., "Technical Summary", in *Climate Change 2013: The physical science basis. Contribution of Working Group I to the Fifth Assessment Report of the Intergovernmental Panel on Climate Change*, ed. T.F Stocker et al. (Cambridge: Cambridge University Press, 2013), 37.

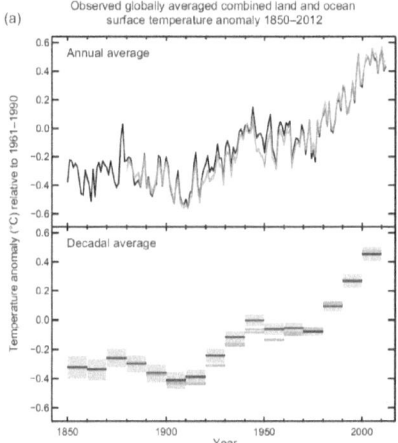

Figure SPM1 from IPCC, 2013: Summary for Policymakers. In: Climate Change 2013: The Physical Science Basis. Contribution of Working Group I to the Fifth Assessment Report of the Intergovernmental Panel on Climate Change [Stocker, T.F., D. Qin, G.-K. Plattner, M. Tignor, S. K. Allen, J. Boschung, A. Nauels, Y. Xia, V. Bex and P.M. Midgley (eds.)]. Cambridge University Press, Cambridge, United Kingdom and New York, NY, USA. (a) Observed global mean combined land and ocean surface temperature anomalies, from 1850 to 2012 from three data sets. Top panel: annual mean values, bottom panel: decadal mean values including the estimate of uncertainty for one dataset (black). Anomalies are relative to the mean of 1961–1990.

The figure above from the *Summary for Policy Makers* of AR5 shows how global average temperatures have changed since 1850, relative to the average temperatures for the period 1961-1990. You can see in the top panel how the three temperature datasets listed earlier are in very close agreement. Large year-to-year variations occur, including the effects of El Niño, which act to cause global temperatures to increase. The bottom panel shows 10 year or decadal averages, and it is very clear that the last decade has been the warmest of the instrumental period.

Recently in the media and the on the internet, people have been talking about a pause or hiatus in the warming. Some say that warming stopped during the 2000s. A careful look at the last figure shows that this has happened before, from the late 1940s to the late 1970s. Does this mean that global warming is all part of a natural cycle? Well firstly, as we have already discussed, the past three decades are warmer than before the Industrial Revolution, with the first ten months of 2014 being the warmest such period on record, 0.68° C above the 20th century average.[50] Things really are getting hotter.

50 NOAA National Climate Data Centre, Global Analysis – October 2014, http://www.ncdc.noaa.gov/sotc/global/, accessed 22 November 2014.

Secondly, we know that volcanoes and human industry can produce particles which can reach high into the atmosphere, reflecting sunlight and cooling the Earth.[51] A number of volcanic eruptions can be tied to cooler years.

Thirdly, the sun varies in its output over time. A recent paper in the journal *Nature* points out that the sun's output has been at a minimum due to the 11-year sunspot cycle.[52] However, these two effects appear not to be enough to explain why the global average temperature rise has not been as rapid as expected. We need to turn to the oceans.

Over 90% of all heat is absorbed by the oceans.[53] Magdalena Balmseda and co-authors in the journal *Geophysical Research Letters* comment that "recent warming rates of the waters below 700m appear to be unprecedented."[54] Something appears to have changed recently so that the oceans are absorbing more heat than before 2000. The likely candidate appears to be changes in the El Niño-Southern Oscillation.[55]

The El Niño-Southern Oscillation is a combination of pressure differences and sea surface temperature patterns across the equatorial Pacific that bring alternating hot and dry (El Niño) and cooler and wet (La Niña) weather to Australia, with the opposite behaviour for the eastern Pacific. It is this phenomenon that gives rise to the famous line from the Dorothea Mackellar poem *My Country* "I love a sunburnt country … Of droughts and flooding rains." During an El Niño, a shift in the location of the largest ocean temperatures from the western Pacific to the central to eastern Pacific means that the oceans release more heat and the global temperatures are warmer overall, by influencing weather patterns in the Northern Hemisphere. During a La Niña, more energy is stored in the oceans.

---

51 The "troposphere" is the region of atmosphere where our weather occurs. It extends from the surface to about 7-8 km in the polar regions and 18 km in the tropics. The "stratosphere" extends from the top of the troposphere to about 50-60 km. The troposphere contains about 99% of all the Earth's water vapour, and hence the stratosphere is very dry and largely cloud-free. It is in the stratosphere where aerosols can reflect large amounts of sunlight.

52 Kosaka Yu, and Xie Shang-Peng. "Recent global-warming hiatus tied to equatorial Pacific surface cooling". *Nature* 501 (28 August 2013): 403-407.

53 Trenberth, Kevin E., and Fasullo, John T. "An apparent hiatus in global warming?," Earth's Future. Doi: 10.1002/2013EF000165.

54 Magdalena Balmaseda, Kevin Trenberth, and Erland Källen. "Distinctive climate signals in reanalysis of global ocean heat content." *Geophysical Research Letters* 40 (16 May 2013): 1754-1759.

55 Yu Kosaka, and Shang-Peng Xie. "Recent global-warming hiatus tied to equatorial Pacific surface cooling". *Nature* 501 (28 August 2013): 403-407.

Since 1998, more heat has been trapped in the oceans, resulting in a decrease in Northern Hemisphere winter temperatures, while at the same time in 2013 Australia had its hottest year on record.[56]

As an important aside, both natural variability and changes in atmospheric circulations due to global warming can lead to some places becoming colder in some seasons in the short term. This reminds us that every time there is a heavy snowfall event in the USA for example, we can't then say "global warming is a hoax." Arctic amplification is the increase in sensitivity to warming or cooling for high northern latitudes.[57] A weakening of the north-south temperature gradient due to strong Arctic warming both weakens the jet stream and makes it wavier. This increases the dragging down of cold Arctic air over North America and Western Europe. This led to extended snow storms in 2009/2010, 2010/2011 and 2012/2013.

The upshot of this is that while the atmosphere and ocean system has regular, internal changes that affect global average temperatures, increased emissions of greenhouse gases mean still more heat to warm up the entire planet, and not just the air we measure with our thermometers.

It is possible to go beyond the instrumental period, using more indirect means than thermometers. When I was living in New South Wales, my automobile insurer floated and went on the stock exchange. As a result, I received a small number of shares. I never make the annual general meeting, but I can always vote indirectly via a proxy form, as if I were there. Climate change proxies work in a similar fashion. We reconstruct the temperature by indirect means, using physical and biological processes that can be calibrated. These include the width of tree rings, the banding in corals, the relative amounts of different isotopes of oxygen in ice cores (oxygen atoms of different weights), sediment cores from lakes and the various pollen types contained, and so on.

Michael Mann was one of the first to compile tree ring records to form what has come to be known as "the Hockey Stick", because the relatively flat temperatures from about 700AD to about 1850, with a sharp upturn after that, resembles a

---

56 Climate Council, *Off the charts: 2013 was Australia's hottest year*, (Melbourne: Climate Council: 2014), https://www.climatecouncil.org.au/ accessed February 3 2014.

57 J.A. Francis. "Rapid Arctic warming and wacky weather: Are they linked?" *American Association for Women in Science Magazine*, Vol. 44. Winter 2013.

hockey stick. Only trees from environments where the major influence of annual growth is temperature can be used for such analysis. Bristlecone pines live at high elevation and in low rainfall climates, and so are ideal for these studies. Mann became the target of much vitriol, both scientifically and personally. Several reviews of his work have been conducted, with different datasets using different types of data such as coral cores, bore holes and glacier records. They all found his hypothesis to be essentially correct.[58] The Fifth Assessment Report notes that this is likely the warmest 30 years for the past 1,400 years. Again, this is the principle of multiple lines of evidence leading to a conviction. If we go back further using other lines of evidence like ice cores, it becomes apparent that the 20th century has reversed a 5,000 year cooling trend in the Northern Hemisphere. It looks like we've gotten ourselves into hot water!

This hot water is also shown in what is happening to sea ice in the Arctic. Arctic sea ice loss has been steadily occurring for some time. Satellite observations show it has been in decline since 1979 when observations first began. Observations from the Danish Meteorological Institute and Norwegian Polar Institute and ocean vessels have shown sea ice cover is much lower than it has been since 1870. Studies of older climate from tree rings, ice cores and lake sediments can take us back even further, showing sea ice is at its lowest for over 1,400 years.

Perhaps the best way to think about sea ice loss is to compare blocks of cheddar cheese to cheese slices. If you want to cut a slice off a block of cheddar, you cut it length ways, with each slice of equal thickness. With each slice you cut off, the surface area of cheese decreases as well as the volume to cheese. Compare that to cheese slices. Each time you remove a slice, the surface area remains the same but the thickness of the cheese decreases, as does the total volume of cheese.

The melting of sea ice is like the removal of cheese slices, except sea ice experiences seasonal variations in thickness. The thickness of ice reflects its age, with the thinnest ice being last season's freezing, and the thickest being from multiple years of freezing. Each summer, sea ice thins due to warming temperatures. Because much of the ice consists of 'thin slices', it is comparatively easy to rapidly lose or gain large areas of ice without greatly increasing the total ice volume. Due to warmer air and ocean temperatures, the total ice volume and hence area has been decreasing over time.

58 A diagram showing how the various studies after Mann's work have confirmed his basic results can be found in Figure 5.7 of chapter 5 of the IPCC AR5.

Over shorter timescales, individual weather events can dramatically affect the total surface area of ice as thin ice area can change rapidly. For example, in 2012, a low pressure system spent nearly two weeks churning up ice already thin from a warm start to the season, which resulted in the lowest sea ice cover on record (though scientists think the ice would have reached a record low anyway). The so-called recovery in 2013 simply means that low pressure systems have brought snow fall and not damaged the ice, resulting in the 6th lowest sea ice cover on record.[59]

Apart from concern over polar bears – that do in fact matter to God, as noted in Section One – Arctic sea ice reflects sunlight back into space, and acts like a refrigerator. To lose the sea ice means to lose one way in which the Earth keeps cool. The loss of Arctic sea ice can reach a tipping point, where the climate alters more or less irreversibly. Climatologist Tim Lenton sets the threshold for this at 0.5-2°C of global warming, so it may be the case we are already committed to losing Arctic sea ice during the summer.[60] Many fossil fuel companies are lining up to make the most of this opportunity with new plans to drill in this region, which will further add to the feedback loop that is global warming!

2.3 Clear and present danger

Key points:

Climate change is happening now: we live in a new climate.

Climate change has increased the occurence and intensity of heat waves, and people have died as a result.

Sea level rise is already impacting the lives and economies of people in the Pacific.

Malaria is spreading due to rising temperatures.

---

59 National Snow & Ice Data Centre (DSIDC), http://nsidc.org/arcticseaicenews/, accessed 12 December 2013.
60 Tim Lenton. "Tipping Elements in the Earth's Climate System," *PNAS* 105 (February 12 2008), 1786-1793.

One interesting aspect of the climate change debate is how we can often distance ourselves from it. A paper in the journal *Global Environment Change* in 2013 entitled *American evangelicals and global warming* suggested that this group of people did not perceive the risk of climate change very highly for themselves or their community.[61] However, amongst them there was a sense of climate change being a risk for future generations and developing countries. Combine this with natural variability, and the difficult and sensitive issue of attributing individual events to climate change, and the whole issue can become a complicated and distant one. It is therefore imperative that we realise that our climate is changing *now*, extreme weather events are becoming *more frequent* (as I will discuss), and that people's lives are being affected, *now!* We have already seen that the world is indeed warming. Now we need to look at what kind of impact this is having on weather, and consequently, human beings.

In a 2012 paper entitled *Framing the way to relate climate extremes to climate change*, noted climatologist Kevin Trenberth states that:

> The answer to the oft-asked question of whether an event is caused by climate change is that it is the wrong question. All weather events are affected by climate change because the environment in which they occur is warmer and moister than it used to be.[62]

So, given temperatures have increased and the atmosphere as a result contains more moisture, we are living in a new climate, with a new normal. This means all weather events are part of this new climate. This is not to say that natural variability has ceased to be, but that, as Trenberth notes, human-induced changes are persistent and in one general direction. Temperature is a good example of this, so let's look at the idea of temperature extremes, and their relationship to a warming world.

---

61   N. Smith and A. Leiserowitz, "American evangelicals and global warming," Global Environmental Change 23:5 (October 2013), 1009-1017

62   Kevin Trenberth, "Framing the way to relate climate extremes to climate change," *Climatic Change* 115:2 (November 2012), 283-290.

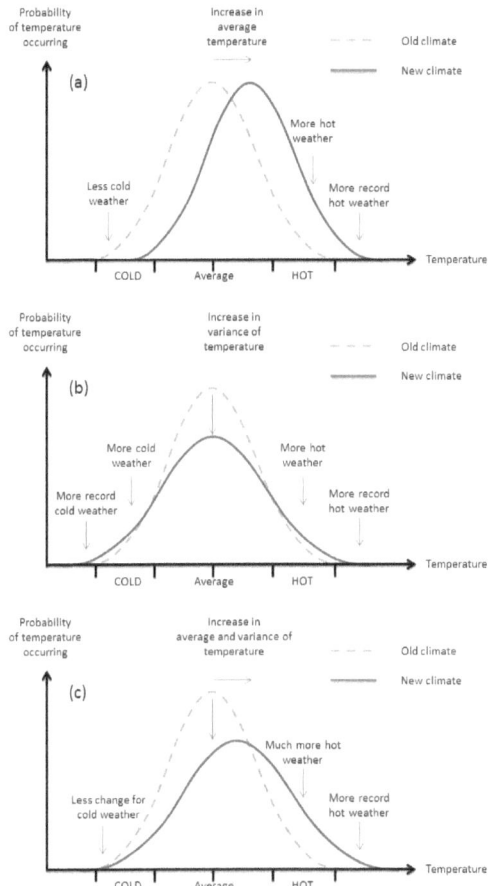

Schematic diagrams showing the effects on extreme temperatures when (a) the mean increases, leading to more record hot weather, (b) the variance increases, and (c) when both the mean and variance increase, leading to much more record hot weather.

The figure above shows that temperature, be it local or global, can be represented by what is known as a normal distribution. Normal distributions are common in a variety of areas and, for example, can be used to plot weights, heights and IQs across a population. For a given value along the horizontal axis, the probability of that value is given on the vertical axis. We can see for the top figure (labelled a) that the most probable value is the average temperature and is shown by the peak of the bell curve. At either edge of the bell curve are extremes, both hot and cold. The probability of extremes occurring is small, hence the small value shown.

Climate change can modify the temperature curve in three ways. In figure (a), the whole climate becomes warmer and the bell curve moves to the right, toward warmer values. Even a modest increase in the average temperature means that the probability of hot days and record hot days increases significantly. In figure (b), the average temperature does not change, but the range over which temperatures vary increases, so that both record hot and cold conditions increase in likelihood. Finally, figure (c) shows what happens when both the average and the range increase. There is a dramatic increase in record hot days.

The European heat wave of 2003 resulted in 70,000 deaths according to an article in *Comptes Rendus Biologies* by Jean-Marie Robine and co-authors. An analysis of summer temperatures in Switzerland shows that the average temperature has increased by 0.8°C, comparing the periods 1864-1923 and 1941-2000.[63] An analogy to illustrate the statistical significance of the 2003 summer in Switzerland to the long term average would be to have a genius walk into a room filled with people of average intelligence.[64] This event is a sign that the average temperature is changing, and possibly also the range of temperatures.

In Australia, the number of record hot days has doubled since 1960 and during the 2009 heat wave, 980 heat-related deaths were recorded, 374 deaths more than would have normally occurred. So, not only is warming occurring, it is having a significant impact in the so-called developed world. Climate change is happening, and it is happening to us!

Additionally, rising sea level contributes to coastal erosion, causing atolls in Tuvalu to be abandoned.[65] Salt water intrusion through porous coral has impacted on taro crops.[66] In the low lying coastal areas of Fiji, recent sugar cane

---

63 Jean-Marie Robine et al., "Death toll exceeded 70,000 in Europe during the summer 2003," *Comptes Rendus Biologies* 331:2 (February 2008), 171-178.

64 For those who understand something of statistics, the 2003 summer temperature was 5.4 sigma levels above the 1961-1990 mean.

65 C.A. Moore, "Awash in a rising sea—How global warming is overwhelming the islands of the tropical Pacific," *International Wildfire*, Jan-Feb 2002.

66 Department of Environment, *Tuvalu's National Adaptation Programme of Action*, Ministry of Natural Resources, Environment, Agriculture and Lands. May 2007. unfccc.int/resource/docs/napa/tuv01.pdf, accessed 11 November 2013.

crops have been poor due to increasingly saline conditions.[67] The main island of the Carteret Islands, home to more than 1,500 people, was completely inundated in 2008.[68] Still bodies of water left over were responsible for malarial outbreaks.[69] In 2007, the Carteret islanders decided to initiate a migration program to Papua New Guinea, although progress has been slow. Loss of fresh water due to sea level rise and intrusion of salt water is considered the most serious climate-related risk for people in places like Kiribati. Conservative estimates from AR5 put global sea level rise at 50 to 100cm by the end of the 21st century. Ultimate the higher end projections are as yet not inevitable.

One of the impacts happening now due to climate change is a change in diseases carried by mosquitoes. The lifecycle of both the mosquito and the malarial parasite speed up as temperature increases, and the mosquito's range is also limited by temperature. The World Health Organization's *World Malaria Report 2012 Fact Sheet* noted that in 2010 there were 219 million cases of malaria worldwide and 660,000 deaths.[70] About 90% of these malarial-related deaths occurred in Africa. Thankfully, between 2000 and 2010, malaria mortality rates dropped 26% globally. There is an inextricable link between poverty and malaria, with the highest mortality rates in countries with the highest rates of extreme poverty.

Andrew Githeko is a world-renowned malaria expert who directs the Climate and Human Health Research Institute at the Kenya Medical Research Institute. He grew up as the eldest of five brothers on a tea farm in the foothills of Mount Kenya, at an altitude of about 1,600 metres. When the British first settled in Kenya in the 1800s they thought the sickness they were catching was due to bad air, hence the name "mal-aria". However, they settled themselves in the cooler region of the highlands, where Githeko grew up, because the disease did not occur there.

67 J Gawander, "Impact of climate change on sugar-cane production in Fiji," *WMO Bulletin* 56:1 (January 2007), 34-39.
68 Julie Edwards. "The Carteret Islands: First man-made climate change evacuees still await resettlement." *Pacific Conference of Churches*, November 2010. http://www.pcc.org.fj/docs/Julias%20Cartaret.pdf, last accessed 11 November, 2013.
69 United Nations University. Local solutions on a sinking paradise, Carteret Islands, Papua New Guinea. http://vimeo.com/4177527, last accessed 11 November, 2013.
70 World Health Organisation (WHO). *World Malaria Report 2012 FACT SHEET* 17 December 2012, http://www.who.int/malaria/publications/world_malaria_report_2012/wmr2012_factsheet.pdf, accessed 9 December 2013.

In 2003, Githeko's niece, Elena, came down with a fever, together with a nasty headache. After a couple of days Elena, dehydrated and with a fever, was taken to hospital. Elena was soon diagnosed with cerebral malaria, a type of malaria where the parasites burrow into the fluid that bathes the brain or even the brain itself. In Africa, a child dies of malaria every 30 seconds. Elena was lucky… but she should never have contracted malaria in the first place. The 1970 national atlas declared the region malaria-free, and the locals have none of the genetic traits that make lowlander Kenyans less susceptible to the disease.

Andrew Githeko was stunned that someone from his childhood home had contracted malaria. He was even more stunned when a visiting scientist found malarial mosquitoes near his home village.[71]

In his research, Githeko has been able to show that the spread in mosquito range and malarial infections matches the observed warming in the region, confirmed by the fact that 80% of the glaciers on Mount Kenya have melted.

This pattern is being repeated in tropical highlands around the world, and paints just one picture of some of the very real and devastating consequences of climate change.

---

71 Paul R. Epstein and Dan Ferber, *Changing planet, changing health: How the climate crisis threatens our health and what we can do about it* (University of California: University of California Press, 2012), 35-42.

2.4 Beyond reasonable doubt

Key points:

Greenhouse gases help keep our planet warm enough for life to exist.

Greenhouse gas concentrations are larger than they have been for the past 800,000 years.

Changes in solar output cannot explain why the lower atmosphere (troposphere) has warmed while the upper atmosphere (stratosphere) has not.

Climate models have shown that for every continent and ocean basin, the rises in temperature observed during the 20th and 21st century cannot be due to natural causes alone, but must include the role of greenhouse gases.

Imagine that you are in a court of law, trying a case of murder. You will want to see the case proved 'beyond reasonable doubt'. Short of having witnessed the murder yourself, you need to be convinced in your own mind that the guilt of the accused is the most likely explanation for the evidence before you. And for that, you will want as much evidence as possible. Not only the smoking gun, but witnesses, DNA evidence, a motive, lack of alibi, and so on. A confession would be nice too.

When it comes to explaining climate change, we have multiple lines of evidence to show that the planet is warming (as already discussed), and that this warming is unlike what has happened in the past. In particular, the fingerprints of humanity are all over the evidence. This makes it pretty clear that humans have caused climate change. And, like a murder trial, a conviction should be delivered to avoid further deaths.

In order to start thinking about why the planet is warming, it is important to note that without greenhouse gases, the planet would be too cold for liquid water, and hence for life as we know it. Imagine going to the bank and changing a 100 dollar note into 10 cent coins. The amount of money is the same but the buying power of each unit is greatly reduced. The sun emits radiation largely in what we call the visible part of the spectrum of radiation, the part humans see. When the Earth and its atmosphere absorb that radiation, it then reemits it at a lower energy, namely the infrared. This is what thermal imaging cameras or satellite pictures show. This is like changing the $100 bill into 10c coins. It's the same amount of heat, but in a lower form. The Earth/atmosphere tries to maintain a balance between money in ($100 bills) and money out (in 10c coins).

The key difference between radiation and a bank account as used in this analogy is that there is a law of physics relating both to the radiation energy that is sent back into space, and to the temperature. This is the issue with global warming: we are changing our energy budget, and hence the temperature associated with this changed budget is increasing. The earth is keeping more of the 10c coins than it should be.

Back in 1824, the Frenchman Joseph Fourier showed that without heat being trapped by the atmosphere, the Earth would be too cold for life. Then, the Englishman William Tyndall showed that these heat-absorbing gases include carbon dioxide, methane and water vapour. Greenhouse gases work by absorbing and then reemitting infrared radiation in all directions. Half of the radiation is sent back down, and the other half back up for every interaction. It is a bit like our money example earlier. Consider the heat energy given up by the Earth's surface; we can think of this energy as paying out of an energy budget, say 10c at a time. Because greenhouse gases can trap and the re-emit this energy in all directions, the net result is that some of it will travel back down to the surface, where it acts to warm the air. This is like paying out 10c of your budget, but getting 5c back.[72] Thus greenhouse gases at natural levels are essential for maintaining life on this planet.

The metaphor of a greenhouse has been used in the past because it illustrates very well how heat can be trapped. Sunlight passes through the glass, but once the air is warmed, it cannot cool as rapidly by mixing with the air outside. I like the image of putting blankets on a bed. The more blankets (greenhouse gases), the warmer the bed becomes. Imagine you are in your bedroom, getting cold. If you turn the heater on, the entire room gets warmer. If instead you put on more blankets, only the bed gets warmer by trapping body heat. The room will actually get a little colder, because the heat your body would have lost to the room is trapped under the blankets.

This bedroom example is useful because it shows why the sun is not primarily responsible for global warming. Apart from the fact that in recent times, the sun has been getting dimmer as part of its 11 year cycle, only the lower atmosphere (the troposphere) where all of the greenhouse gases are found, has been getting warmer. The stratosphere has been getting cooler. This is because more heat is trapped by the greenhouse blanket in the troposphere and hence the stratosphere is missing out. If the sun were brightening, both layers would be warm.

Water vapour is the most significant of the greenhouse gases because of its sheer abundance, and this is sometimes used to deny that carbon dioxide is responsible for warming the planet. What we need to do is understand the difference between "forcing", "feedback", and "sensitivity". When you strum an electric guitar, you are forcing the strings to vibrate, and this sound is detected by the guitar pick-up, and the signal is passed to an amplifier. Together, the settings of the pick-up and the amplifier set the sensitivity of the system to the force of the strum. The same strum can produce a different volume, depending on the sensitivity as set by the pick-up and amp. This is why the IPCC report

---

72 The reason that some gases like carbon dioxide and water vapour are able to absorb and reemit heat is their shape. Unlike oxygen and nitrogen, which consist of two atoms with fairly inflexible bonds in between them, water vapour and carbon dioxide involve three atoms. The bonds can bend, absorbing incoming radiation (heat). When the bonds flex back, the energy is then reemitted in all directions, warming the atmosphere.

has a range of warming for doubling atmospheric CO_2 over pre-industrial levels, because we don't know precisely the climate amplifier settings. This is due to a lack of 100% knowledge of all of the feedbacks.

Now take our guitar and put it in front of the amplifier. The sound from the amplifier will be added to the strum and collected by the pick-up. This means that more noise comes out of the amplifier, and so on. This is known as positive feedback, and in the case of the guitar it produces distortion. To relate this to the climate, think about the way in which greenhouse gases are added to the planet. Carbon dioxide and methane can be produced by burning fossil fuels. Carbon dioxide also comes out of volcanoes, and is lost when plants die. Methane comes from cows and sheep, and from rotting organic material. We are *forcing* the climate by adding these gases. But what about water vapour? Water evaporates when temperatures are above freezing, and as temperatures increase, so does the rate of evaporation. Water vapour acts as a *feedback* like the guitar in front of the amp. As temperatures increase due to greenhouse gases, so more water vapour is added to the atmosphere, warming temperatures further, and so on. This total feedback provides us with the uncertainty in the *sensitivity*.

So what evidence do we have for our "murder trial" that these greenhouse gases are increasing? Firstly, we have direct measurements from the late 1950s. The now famous "Keeling curve", named after Charles Keeling, who started recording carbon dioxide concentrations at Mauna Loa in Hawaii, shows an increase in carbon dioxide from about 315 parts per million (ppm) to 400 ppm. This value goes up and down during the year by a few ppm due to the winter dormancy of the Northern Hemisphere forests as they are covered in snow.

Secondly, we can measure greenhouse gases via ice cores. An ice core is a sample of ice removed from an ice sheet, typically from the polar ice caps of Antarctica or Greenland. Ice forms from the incremental build-up of annual layers of snow. The properties of the ice or inclusions can then be used to reconstruct a climatic record. Inclusions in the ice include wind-blown dust, ash, bubbles of atmospheric gas and radioactive substances. Climate data that may be inferred from these inclusions are temperature, ocean volume, precipitation, chemistry and gas composition of the lower atmosphere, volcanic eruptions, solar variability, sea-surface productivity, desert extent and forest fires.

Ice cores show that atmospheric carbon dioxide increases and decreases cyclically over time as ice ages come and go. A recent paper in *Nature* suggests that the oceans' storage and release of carbon dioxide changes with changes in ocean circulation.[73] The changes in temperature have been driven by changes in the amount and way sunlight is distributed about the globe, with carbon dioxide and methane acting as a feedback. What is different about the end of the last ice age some 10,000 years ago is that the concentrations of greenhouse gases have continued to climb, consistent with the burning of fossil fuels. In fact, AR5 states with very high confidence that greenhouse gas concentrations are larger than they have been for the past 800,000 years.

The last piece of evidence to look at in our trial is global climate models. While no one is pretending they contain every last detail of physics or that all uncertainties are nailed down, it is significant that climate models can reproduce the 20th and 21st century changes in temperature only if the changes in greenhouse gases are included. Natural changes to the sun's output and to the concentration of sunlight reflecting aerosols are not enough. The Fifth Assessment Report shows that this is the case not only for the global average, but for every continent and oceanic basin.

---

73 A.N. Meckler, et al. "Deglacial pulses of deep-ocean silicate into the subtropical North Atlantic Ocean," *Nature* 495 (28 March 2013): 495-499.

2.5 Gaze into my crystal ball

Key points:

Carbon dioxide stays in the atmosphere for a very long time.

2°C warming above pre-industrial levels is considered the safety threshold, yet we have already seen profound impacts for less than 1°C of warming.

Global greenhouse gas emissions are consistent with the most pessimistic of all our emissions scenarios and we are well on the way to a very different world.

Unless we make very rapid changes to our emissions, we will not be able to avoid 2°C of warming.

One of the most important key uncertainties in how the future will unfold is how human beings will react. To date, the human response has not been very promising. Scientists have been concerned for over four decades that the planet is warming, and yet little has changed.

In order to capture the uncertainty in how much greenhouse gases humans will produce via fossil fuel burning, land clearing and so on, AR5 uses Representative Concentration Pathways, or RCPs for short, shown in the figure below.[74] These are scenarios showing the likely concentration of greenhouse gases in the atmosphere depending on how much and how quickly humans increase or reduce the rate of greenhouse gas emissions. The RCPs are labelled 2.6, 4.5, 6.0 and 8.5 in order of increasing climate impact.

---

[74] H. Richard, et al. "The next generation of scenarios for climate change research and assessment" *Nature* 463 (11 February 2010): 747-756. The figure comes from chapter 5 of the IPCC WGI AR5.

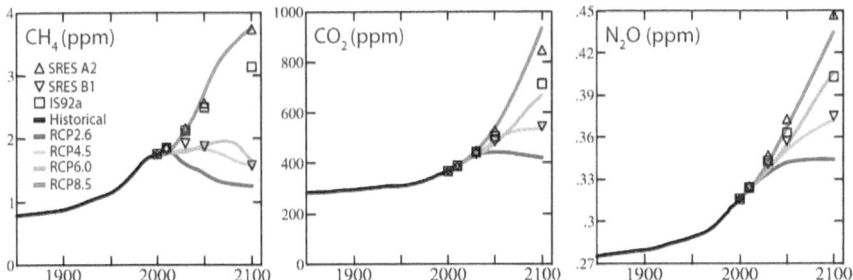

Figure 8.5 from Myhre, G., D. Shindell, F.-M. Bréon, W. Collins, J. Fuglestvedt, J. Huang, D. Koch, J.-F. Lamarque, D. Lee, B. Mendoza, T. Nakajima, A. Robock, G. Stephens, T. Takemura and H. Zhang, 2013: Anthropogenic and Natural Radiative Forcing. In: Climate Change 2013: The Physical Science Basis. Contribution of Working Group I to the Fifth Assessment Report of the Intergovernmental Panel on Climate Change [Stocker, T.F., D. Qin, G.-K. Plattner, M. Tignor, S.K. Allen, J. Boschung, A. Nauels, Y. Xia, V. Bex and P.M. Midgley (eds.)]. Cambridge University Press, Cambridge, United Kingdom and New York, NY, USA.: The figure shows the time evolution from 1850 to 2100 of greenhouse gas concentrations in parts per million by volume for methane (left), carbon dioxide (middle) and nitrous oxide (right). The black line represents historical emissions of each of the gases to 2000, the end of the dataset for the study. The diverging lines represent each of the four RCPs.

The RCPs may be thought of in three ways. Firstly, we can think about them in terms of the amount of greenhouse gases emitted over time. In terms of our analogy of the bed from earlier, the larger the amount of greenhouse gases, the more the number of heat-trapping blankets we will be adding to the atmosphere. Focussing on carbon dioxide, the pre-industrial value was 280 ppm. We are already at 400 ppm. The worst of the RCPs sees the amount of carbon dioxide nearly five times the pre-industrial value by the end of this century. That's a lot of extra blankets and hence a lot of extra energy trapped, and therefore a much warmer planet.

The second way to think about these scenarios is to think about the extra energy trapped by our greenhouse blankets. We can think about the RCPs in terms of an electric bar heater. The higher the value of the RPC, the more bars are switched on and so the more energy you are using to heat the room (or in our case, the atmosphere). To switch metaphors slightly again, based on measurements of the ocean's heat, (the ocean absorbing approx. 93% of the earth's heat), humans are currently adding the equivalent energy of four Hiroshima bombs each second to our Earth system![75]

---

75 See for example http://4hiroshimas.com, accessed 21 January 2014.

Finally, we can think about how each of these RCPs translates into how much warmer the world will be by the end of the century. It is sometimes claimed that 2°C above pre-industrial levels is a safe threshold over which we must not pass. Only the lowest emissions scenario will keep us under 2°C, and this scenario is associated with aggressive cutting of greenhouse gas emissions, far more aggressive than any nation is achieving right now. While 2°C of warming doesn't seem like much, remember a few things. Firstly, we've already seen dramatic changes around the globe in terms of spread of disease, increased drought, heat waves and bush fires, as well as an increase in significant storms and floods, and all with changes of less than 1°C above pre-industrial levels.

Secondly, if we also take a longer term view, a world warmed by 2°C will be "decidedly warmer than the Earth has been in millions of years" and would remain at least 1°C warmer for more than 10,000 years![76] This is not simply climate change but climate disruption. The changes we have already made are geologically significant, and will outlast not only our lifetimes, but those of many civilisations.

A 2°C warming has been associated in the geologic past with sea levels some 20 metres above present levels, but clearly this did not happen overnight – it takes time to melt ice sheets! Climate projections usually are discussed in terms of sea level rise of 10s of centimetres or a metre or so within a hundred years. But sea level rise could be more rapid than anticipated. The conservative high-end warming scenarios project a sea level rise of about 80cm by end of century[77], however some modelling approaches project double that figure., Uncertainty exists however in modelling Antarctic ice loss, with a recent study showing a 77%

---

76 David Archer, and Victor Brovkin, "The millennial atmospheric lifetime of anthropogenic CO2." *Climatic Change* 90 (2008): 283-297.

77 Church, J.A., P.U. Clark, A. Cazenave, J.M. Gregory, S. Jevrejeva, A. Levermann, M.A. Merrifield, G.A. Milne, R.S. Nerem, P.D. Nunn, A.J. Payne, W.T. Pfeffer, D. Stammer and A.S. Unnikrishnan, 2013: Sea Level Change. In: Climate Change 2013: The Physical Science Basis. Contribution of Working Group I to the Fifth Assessment Report of the Intergovernmental Panel on Climate Change [Stocker, T.F., D. Qin, G.-K. Plattner, M. Tignor, S.K. Allen, J. Boschung, A. Nauels, Y. Xia, V. Bex and P.M. Midgley (eds.)]. Cambridge University Press, Cambridge, United Kingdom and New York, NY, USA, 1140.

increase in ice loss from area since 1973.[78] Over the longer term, contributions from expanding sea water and melting ice sheets should contribute about 2.3m of sea level rise per degree Celsius of warming over the next 2,000 years.[79]

James Hansen and co-authors note that if we look back in time, during the Eemian (between 130,000 and 114,000 years ago), conditions were 2°C warmer than the pre-industrial revolution[80.] Sea level was some 9m above present. This is because there was plenty of time to melt the continental ice sheets such as East Antarctica. Projections of a metre or so of sea level rise by the end of the century might seem okay to some, but sooner or later (and we don't know how much sooner) 2°C of warmer weather will likely cause the mass migration of billions of people. Will there be room for all of them?

Human displacement. Climate change. Two of the world's leading causes of concern for humanitarians, environmentalists and the public at large alike. But what is even more concerning is that the two are inextricably linked – a growing global problem that researchers have now coined 'climate displacement'.

According to the United Nations' High Commissioner for Refugees, the global scale of climate displacement is expected to dwarf today's levels of displacement, which UNHCR recently announced are higher than at any time since the end of the Second World War.[81]

With this in mind, a commentary by Glen Peters and co-authors in *Nature Climate Change* highlights how difficult it is going to be to keep warning to below 2°C.[82] Historical emissions of greenhouse gases to 2011 are above those

---

[78] J Mouginot, E Rignot, and B. Scheuchi, "Sustained increase in ice discharge form the Amundsen Sea Embayment, West Antarctica, from 1973 to 2013", *Geophysical Research Letters* 41 (16 March 2014): 1576-1584.

[79] Anders Levermann, et al. "The multimillenial sea-level commitment of global warming" *Proccedings of the National Academy of Science* 110 (August 20, 2013): 13754-13750.

[80] James Hansen, et al. "Assessing "Dangerous Climate Change": Required reduction of carbon emissions to protect young people, future generations and nature". *PLOS* One 8 (December 2013). http://www.plos.org/publications/journals/, accessed October 2014.

[81] Scott Leckie, "Preparing for the climate displaced, both rich and poor" *Climate Spectator* (online), 24 July 2014, http://www.businessspectator.com.au/article/2014/7/24/policy-politics/preparing-climate-displaced-both-rich-and-poor? accessed 20 October 2014.

[82] Glenn Peters, et al. "The challenge to keep global warming below 2° C." *Nature Climate Change 3* (January 2013): 4-6.

for RCP8.5, which is associated with an end-of-century temperature range of 3.5 to 6.2°C. Emissions have grown by 3% on average since 2000. According to Peters, to avoid exceeding 2°C will require "high levels of technological, social and political innovations, and an increasing need to rely on net negative emissions in the future."

"Net negative emissions" means either the effective carbon capture and storage from the combustion of fossil fuels, or the direct removal from the atmosphere of carbon dioxide. The former has promised much and delivered little. Hansen and co-authors (2013) note that: "At present there are no proven technologies capable of large-scale air capture of CO_2". To avoid having to rely upon very expensive technologies (some of which are still not beyond the drawing board) means starting now with dramatic cuts in our emissions. Staying under 2°C means emitting no more than 565 gigatons of CO_2, while *The Carbon Tracker* initiative suggests that there is five times that much stored in fossil fuel reserves.[83]

In the Australian context, the plan to open new coal ports in the Great Barrier Reef, a beautiful world heritage listed area and popular tourist destination, makes neither economic or environmental sense, when this stunning example of God's creation is already unlikely to survive a 2°C world. Nor does it make sense for state governments to consider exporting brown coal (which emits more CO_2 than black coal), or continue mining for coal seam gas or shale oil.

Hansen and co-authors suggest that 2°C is not only too warm, but that slow feedbacks such as the release of methane from melting tundra (adding to the atmospheric greenhouse gases) and melting ice sheets (again the loss of large areas cooling the planet by reflecting sunlight back into space) would inevitably result in 3°C of warming instead.

If 3°C is dramatic enough, current emissions are taking us on the way to 4°C of warming. In September of 2009, a meeting was held in the UK called *4 Degrees and Beyond*. An article in *Nature Reports Climate Change* in 2009 summarises the horror that is a 4°C world (warmer than the past 800,000 years). The permanent breakup of the Greenland ice sheet would add to sea level rise by about 7m. An ice-free summer in the Arctic would see the end of species like polar bears,

---

83 Figures obtained from "Fossil Free" website: http://gofossilfree.org/faq/ accessed 29 November 2013.

and the Earth would lose its cooling effects on climate. Coral reefs would be heat-stressed out of existence or drowned by lack of sufficient sunlight, and with them we would lose a major source of tourist dollars as well as protein for many island nations (to say nothing of the drowning of the islands themselves!) Higher temperatures in northern Australia and the Sahara would be recorded than anywhere in the world today. More than 80% of land area in most African nations would be unsuitable for crops (at 2°C warming), resulting in the deaths of millions to starvation. In the most currently water-stressed locations around the world, 4°C would reduce clean water availability for 50% of residents.[84]

Perhaps most chillingly, human population would be devastated by 4°C of warming. The report *4 degrees hotter*, produced by the Climate Action Centre, quotes Professor Kevin Anderson, director of the Tyndall Centre for Climate Change as suggesting that only 10% of humanity would survive![85] Increased intensity of bush fires, droughts, storms and floods will combine with sea-level rise to create a world very unlike the one we know now. Disaster and emergency will be the new normal. Survival will become the order of the day, with extremely limited opportunities for the kind of 'human flourishing' we have become accustomed to.

A final word is needed regarding all of these projections: all of the temperatures quoted refer to the end of the century. So long as we continue to burn fossil fuels, the planet will only get warmer, potentially resulting in explosive feedbacks. About 250 million years ago, warming of the ocean surface to temperatures up to 40°C due to volcanic eruptions led to the largest mass extinction in history.[86] It remains to be seen whether or not human activity would result in such large feedbacks such as a 'methane burp' from oceanic deposits,[87] but we have already

---

[84] Mark New, Diana Liverman, and Kevin Anderson. "Mind the gap". Nature Reports: *Climate Change 3.* (December 2009): 143-144.

[85] Climate Action Centre, 4 degrees hotter: *A climate action primer* (Melbourne: Climate Action Centre, 2011). http://fisnua.com/wp-content/uploads/2011/03/4-degrees-hotter.pdf, accessed 11 February 2014.

[86] Charles Choi, "Permian-Triassic extinction event may have been driven by extreme warming." *HuffPost Green*, 18 October 2012. http://www.huffingtonpost.com/2012/10/18/permiantriassic-extinction-event_n_1981835.html, accessed 13 February 2014.

[87] Tim Lenton, "Tipping elements in the Earth's climate system," *PNAS* 105:6 (February 12 2008), 1786-1793, suggests the temperature threshold for the release of marine methane hydrates is currently unknown. Scientists still don't agree.

seen how ignorance (or uncertainty) is no excuse for inaction. We already have the potential to devastate the planet and ourselves – in fact we are on track for exactly that. There may be no limit to our destruction. In the absence of such knowledge we should be acting to avoid any serious outcomes, and acting now.

It is time to turn our economies and our societies around before we head over the precipice toward a warming world. We have only just begun to see the impacts a small amount of warming has. It is still possible for us to limit warming and its impacts, but the more we delay, the harder and more costly it will be. Governments agreed this needed to be done back in Rio at the Earth Summit in 1992, but have done little since. What this turn around, this change of mind (otherwise known as repentance) looks like, is the subject of Sections Four and Five.

Section Three: Understanding the times

Claire Dawson

3.1 Cool, calm and collected

Key Points:

Climate Change poses a massive threat to society. This is very bad news!

We have now known about many of these risks, with increasing certainty, for decades.

Our personal context significantly shapes our response to the issue.

There are many reasons for our inadequate responses to the challenge of climate change, including a profound lack of leadership within broader society.

So, by now you've worked your way through some reasonably heavy theology and science. If you're beginning to think that we have a very serious problem on our hands then you're getting the picture. Realising the gravity of our current predicament, you may well be wondering "How on earth did we get *here*?" The window of opportunity for genuine, effective action is now disturbingly small, and the hard reality is that we have failed to stop climate change in its tracks: it is *already happening*.

This issue may not have been urgent when academics began focussing more attention on it in the late 1970's, but it certainly is now. Some even suggest that we're now well past the place where various tipping points will inevitably lead to reasonably catastrophic changes in our global climate, regardless of what we do next.

In climate change circles one occasionally encounters people who subscribe to views under the broad description of "NTHE" (Near Term Human Extinction). These people are convinced that human society does not have what it takes to change within the necessary timeframe, and feel certain that we are among the last people to experience human civilisation as we know it. Some even see the extinction of the human species as the only thing that will give our planet some breathing space – after which time it can slowly recover and regenerate!

The most concerning aspect of such views is that there is generally *no hope*, and no reason to act. Not surprisingly, one doesn't hear much from those in the NTHE camp as they don't see any point in holding conferences, producing media releases or creating fancy websites to convert others to their point of view. It is all so futile. There is no point. This must be a debilitating view to hold.

While Mick and I certainly don't share these more extreme views, we are willing to acknowledge that in our era, *homo sapiens* have demonstrated a tendency to avoid bad news – especially news that demands a significant response from us. It is constructive to contemplate what is actually at stake here, just how urgent the situation is now, and to try to slowly cut through the layers of denial and unbelief. The fact is, we have known about this prognosis for decades, and have until very recently failed to act at almost every level. People are suffering now as a result, and collectively we are culpable.

I remember learning about 'global warming' in geography class back in the late 1980s. Even then things seemed so bad to me that I felt quite convinced that I would never choose to have children of my own. The future seemed bleak, and humanity seemed to be failing miserably in its responsibility to care for the planet. I remember flying into Los Angeles as a 16 year-old and being completely heart-broken by the thick layer of brown smog that clouded the horizon. In my young mind the USA was the pinnacle of human achievement, yet what price was being paid for all of this progress and development? At the NASA space centre the message of the day seemed to be that the hunt was on for other habitable planets – because we'd made such a mess of our own!

At the time I was without faith, and most certainly without hope, and it was these big questions about life – and my growing pessimism about humanity's future – that actually fuelled my spiritual search. My journey included a pivotal turning point where I embraced an evangelical expression of the Christian faith toward the end of my final year at school. Within the church circles I was introduced to, however, environmental concern was well and truly off the radar. Tree-hugging 'liberals' were sometimes publicly belittled by church leaders. 'Personal holiness', 'evangelism' and 'mission' – in the most narrow senses of those terms – seemed to be the only things that mattered in discipleship, at the exclusion of all other concerns.

I learnt a bit more about global warming at university while studying economics. I learnt that air pollution is a classic example of an 'externality' and that externalities occur when the price of a product or service doesn't factor in the full social (or environmental) cost. This was an example of 'market failure'. While we needed to know this as theory, there was no strong sense that rising levels of carbon dioxide emissions was a particularly serious or urgent planetary problem. Our discussion was an academic exercise: we needed to know and accept that our current economic model (market capitalism) was not quite perfect. Fortunately by the mid 1990s people within the Faculty of Science were beginning to take the threat of global warming much more seriously than those within the Faculty of Business and Economics!

Over the past two decades my views regarding environmental concern have certainly been changing and evolving, and my involvement with church has certainly been a significant influence. At worst, I developed such an other-worldly faith that earthly issues were quickly and easily written off as a distraction. At best, the call of Christ to live with love, justice and compassion has compelled me to seriously engage with this issue. And this journey of discovery has been enhanced and enriched by the inspiration of some amazingly gifted and committed Christians – as well as some extremely compassionate and dedicated people who reside well outside of the walls of the established Church.

As my own story demonstrates, our own life stories obviously play a significant part in shaping our values, attitudes, and ethical behaviour. We are in many ways products of our various contexts, including families, friendship circles, churches, schools/universities, workplaces and community groups. And we are all influenced, to at least some extent, by the views of broader society. While dwelling on the past is not always constructive, I think in this case it's important to understand a little bit about the reasonably complex interplay of issues and powers at work in the 'climate change debate' in recent decades – often very much behind the scenes. It is far from straightforward, and ultimately it would seem that in a number of ways climate change is a symptom of far deeper issues relating to our addiction to our materialistic way of life.

Looking back at how things have unfolded, it quickly becomes clear that time and time again we have consistently failed to act reasonably and effectively in response to the threat of climate change. In the longer-term scheme of things we have not acted in our own best interests, or those of our neighbours or our grandchildren. Why is this? Most of us would not choose to drive over a bridge that 97% of civil engineers had deemed unsafe, and most of us would not take a medication or undergo a medical procedure that 97% of medical professionals deemed to be very hazardous to human health… and yet when it comes to climate change, we have somehow remained cool, calm and collected while ignoring the expert opinions and increasingly urgent warnings of at least 97% of the scientific community!

There are a range of reasons why climate change has elicited such an irrational response from the general population. We will explore a few of these reasons in more detail, but one could summarise broadly to say that there has been a *profound lack of leadership* with regard to this crucial issue. Additionally, where people and institutions have taken the lead and managed to influence outcomes, unfortunately their agenda has generally been to hinder effective action for short-term gain (financial or political).

So now we need all the wisdom and inspiration we can muster in order for us to engage ourselves in efforts to turn this situation around. We need to comprehend what has been happening in recent decades so that we can better understand the times we are in, and discern what we must do in response.

Savouring the last Federal election win.

3.2 Prophecy & Pride

Key Points

God generally warns us when we are on the wrong path, through Scripture, through his people, and sometimes through other means.

The Bible offers clear warnings against injustice, idolatry, greed and excess.

Over the past forty years there have been specific warnings issued about ecological damage, and climate change more specifically, including warnings from faithful Christians.

The church has missed a vital opportunity to demonstrate radical faithfulness and love, and to lead the world in bringing positive and necessary change.

We are already suffering the consequences of a disrupted climate.

For many Christians, the issue of climate change can be one that causes unease. It is indeed a massive issue, and everyone should actually feel unease about it: that would in fact be a very healthy place to start! Yet our discipleship journey does often not prepare us well for grappling with such things – especially when the Bible appears – at least superficially – to say very little about it.

Mick has already addressed some of the specific theological concepts that help us to better 'frame' the issue of climate change within a biblical context. There will be a bit of overlap here as I revisit some similar themes: I hope it will serve to reinforce and amplify our core message.

"Didn't I tell you so?"

Time and time again, the Scriptures introduce us to a righteous and compassionate God who expresses his care for people by issuing warnings via his prophets. At Mount Sinai God established his Covenant with Israel, and promised blessings in response to their continued faithfulness to him and the way of life he called them to.

> Now what I am commanding you today is not too difficult for you or beyond your reach. It is not up in heaven, so that you have to ask, "Who will ascend into heaven to get it and proclaim it to us so we may obey it?" Nor is it beyond the sea, so that you have to ask, "Who will cross the sea to get it and proclaim it to us so we may obey it?" No, the word is very near you; it is in your mouth and in your heart so you may obey it. See, I set before you today life and prosperity, death and destruction. For I command you today to love the Lord your God, to walk in obedience to him, and to keep his commands, decrees and laws; then you will live and increase, and the Lord your God will bless you in the land you are entering to possess.
>
> But if your heart turns away and you are not obedient, and if you are drawn away to bow down to other gods and worship them, I declare to you this day that you will certainly be destroyed. You will not live long in the land you are crossing the Jordan to enter and possess.
>
> This day I call the heavens and the earth as witnesses against you that I have set before you life and death, blessings and curses. Now choose life, so that you and your children may live and that you may love the Lord your God, listen to his voice, and hold fast to him. (Deuteronomy 30:11-20a)

From that time on it seems that God has been busy calling his people back to himself, back to covenant relationship with him. Time and time again he appeals for his people to turn from their idolatry and sin, making appeals through his messengers, the prophets.

You have forgotten God your Savior;
> you have not remembered the Rock, your fortress.
Therefore, though you set out the finest plants
> and plant imported vines,
though on the day you set them out, you make them grow,
> and on the morning when you plant them, you bring them to bud,
yet the harvest will be as nothing
> in the day of disease and incurable pain. (Isaiah 17:10-12)

See, the Lord is going to lay waste the earth
> and devastate it;
he will ruin its face
> and scatter its inhabitants—
it will be the same
> for priest as for people,
> for the master as for his servant,
> for the mistress as for her servant,
> for seller as for buyer,
> for borrower as for lender,
> for debtor as for creditor.
The earth will be completely laid waste
> and totally plundered.
The Lord has spoken this word.

The earth dries up and withers,
> the world languishes and withers,
> the heavens languish with the earth.
The earth is defiled by its people;
> they have disobeyed the laws,
violated the statutes
> and broken the everlasting covenant.

Therefore a curse consumes the earth;
> its people must bear their guilt.
Therefore earth's inhabitants are burned up,
> and very few are left.
The new wine dries up and the vine withers;
> all the merrymakers groan.

The joyful timbrels are stilled,
   the noise of the revellers has stopped,
   the joyful harp is silent.
No longer do they drink wine with a song;
   the beer is bitter to its drinkers.

The ruined city lies desolate;
   the entrance to every house is barred. (Isaiah 24)

You women who are so complacent, rise up and listen to me; you daughters who feel secure, hear what I have to say! In little more than a year you who feel secure will tremble; the grape harvest will fail, and the harvest of fruit will not come. (Isaiah 32:9-10)

This is what the Lord says:
"Stand at the crossroads and look;
   ask for the ancient paths,
ask where the good way is, and walk in it,
   and you will find rest for your souls.
   But you said, 'We will not walk in it.'

I appointed watchmen over you and said,
   'Listen to the sound of the trumpet!'
   But you said, 'We will not listen.'
Therefore hear, you nations;
   you who are witnesses,
   observe what will happen to them.
Hear, you earth:
   I am bringing disaster on this people,
   the fruit of their schemes,
because they have not listened to my words
   and have rejected my law.
What do I care about incense from Sheba
   or sweet calamus from a distant land?
Your burnt offerings are not acceptable;
   your sacrifices do not please me."

Therefore this is what the Lord says:
"I will put obstacles before this people.
  Parents and children alike will stumble over them;
  neighbors and friends will perish." (Jeremiah 6:16-21)

The imagery of standing at the crossroads and needing to discern to right path is of course very pertinent. In his love God gives us freedom: freedom to choose. Will we choose the right path, the path that leads to life? Or will we choose the wrong path, the path that leads to death?

Seek good, not evil,
  that you may live.
Then the Lord God Almighty will be with you,
  just as you say he is.
Hate evil, love good;
  maintain justice in the courts.
Perhaps the Lord God Almighty will have mercy
  on the remnant of Joseph. (Amos 5:14-15)

Loving good and maintaining justice are part of keeping covenant. God calls us to love him through loving others. This involves sacrifice and effort, the antithesis of which is selfish indulgence and laziness. In light of this, these words from Amos also seem profoundly apt:

You put off the day of disaster
  and bring near a reign of terror.
You lie on beds adorned with ivory
  and lounge on your couches.
You dine on choice lambs
  and fattened calves.
You strum away on your harps like David
  and improvise on musical instruments.
You drink wine by the bowlful
  and use the finest lotions,
  but you do not grieve over the ruin of Joseph.
Therefore you will be among the first to go into exile;
  your feasting and lounging will end. (Amos 6:4-7)

For their sin, for their failure to keep covenant, God's people did indeed spend time in exile, far off in Babylon. They lost their land: their very precious inheritance. Centuries after the earlier prophecies of Amos and Isaiah, God reminds his People of why all of this happened, through the prophet Zechariah:

> And the word of the Lord came again to Zechariah: "This is what the Lord Almighty said: 'Administer true justice; show mercy and compassion to one another. Do not oppress the widow or the fatherless, the foreigner or the poor. Do not plot evil against each other.'
>
> "But they refused to pay attention; stubbornly they turned their backs and covered their ears. They made their hearts as hard as flint and would not listen to the law or to the words that the Lord Almighty had sent by his Spirit through the earlier prophets. So the Lord Almighty was very angry.
>
> "'When I called, they did not listen; so when they called, I would not listen,' says the Lord Almighty. 'I scattered them with a whirlwind among all the nations, where they were strangers. The land they left behind them was so desolate that no one traveled through it. This is how they made the pleasant land desolate.'" (Zechariah 7:8-13)

As Christians, we revere the Old Testament as being God's very words to us. We allow God to speak to us through it, and we learn from its richness. However we take care to interpret it through the lens of the New Testament and the New Covenant established in Jesus Christ, which is better and superior to the Old Covenant. Yet, not surprisingly, the call is still to extend love:

> A new command I give you: Love one another. As I have loved you, so you must love one another. By this everyone will know that you are my disciples, if you love one another. (John 13:34-35)

And while we so often continue to miss the mark, we take great hope in the promise of forgiveness and grace through Jesus Christ, which is at the heart of this new covenant:

> Then he took a cup, and when he had given thanks, he gave it to them, saying, "Drink from it, all of you. This is my blood of the covenant, which is poured out for many for the forgiveness of sins." (Matthew 26:27-28)

Accordingly we do not live in fear of judgement, wrath or the punishment of death, as people did under the Old Covenant. Instead we are to use our freedom to pursue lives characterised by love and obedience. I felt it important to reflect on the New Covenant in order to establish that we do not to respond to God out of legalism and fear. Rather, we respond out of faith in Christ, out of thankfulness for all that he has done, being energised by the hope held out to us in the resurrection of Jesus. Indeed, this New Covenant changes so much, yet the way of life he has established for us, his desire that we live in accordance with his ways – seeking justice and extending mercy – is still at the heart of loving relationship with him. It is love that should characterise us as people and as communities.

Interestingly, the prophecy of Agabus as recorded in the New Testament predicts a famine that would impact the entire Roman world. This was in an era when the church was new, still finding its feet, so to speak, and growing by the day:

> During this time some prophets came down from Jerusalem to Antioch. One of them, named Agabus, stood up and through the Spirit predicted that a severe famine would spread over the entire Roman world. (This happened during the reign of Claudius.) The disciples, as each one was able, decided to provide help for the brothers and sisters living in Judea. This they did, sending their gift to the elders by Barnabas and Saul. (Acts 11:27–30)

In this particular case we do not read about the disciples responding with prayer for God to halt the famine. Neither do we read that they repented of any specific sins which were the apparent 'cause'. Rather, those who were able dug deep financially in order to provide for those in need. Their response was purely practical, one of service and love. In fact, this is credited as being the earliest recorded example of a Christian 'relief project' – and an international one at that![88] This is a particularly relevant text for us, as a globally just response to climate change has implications for our commitment to support – and indeed increase – our levels of compassionate giving and overseas aid.

When it comes to climate change, if we believe that God cares about his good creation and longs for it to flourish and remain habitable, and if it is the poor who will suffer first, and if he has expressed a particular concern for those who are most vulnerable, then surely it would be reasonable to expect some warnings if we were heading down the wrong path? And surely God's passion for justice would be ignited if it is the selfish, indulgent and excessive lifestyles of the few that will cause misery and oppression for the many?

Those within the evangelical tradition hold strongly to the view that God speaks primarily through Scripture, and there is no shortage of warnings against greed and excess and the danger that awaits those who pursue selfish gain. These verses from James offer an appropriate example:

> Now listen, you rich people, weep and wail because of the misery that is coming upon you. Your wealth has rotted, and moths have eaten your clothes. Your gold and silver are corroded. Their corrosion will testify against you and east your flesh like fire. You have hoarded wealth in the last days. Look! The wages you failed to pay the workmen who mowed your fields are crying out against you. The cries of the harvesters have reached the ears of the Lord Almighty. You have lived on earth in luxury and self-indulgence. You have fattened yourselves in the day of slaughter. You have condemned and murdered innocent men, who were not opposing you. (James 5:1-6)

We have certainly hoarded wealth. Undoubtedly for many people wealth is exists in the form of various financial assets, but in our era particularly it also exists in the form of more and more 'stuff': we are surrounded by so much stuff that it has become normal, and hardly noticeable! We are slowly becoming aware of how our cheap products often come at the expense of others: people whose land

---

88 John Dickson refers to this in World Vision's DVD series, *The Faith Effect* (Session 2 - Religion Poisons Everything), 2012.

has been despoiled and people whose toil is neither safe, dignified, or adequately compensated (including the work of mere children and those in bonded labour).

The ugly underbelly of our consumer economies in the West has gorged on the lives of the poor in the majority poor world – all so that our insatiable desires for more cheap stuff can be fed again and again! Unlike other environmental disasters, climate change is truly global in its reach, so in a very real sense our actions (and our inaction) have collectively condemned others to live with radically destabilised climates, even though many of these people live incredibly simple lives. Thus, many of those who suffer now have done next to nothing to cause this global environmental crisis. Reflecting upon these challenging words from the book of James, we may very well be contributing to the death of innocent people, who have in no way been opposing us!

In drawing upon Scripture to reflect on climate change, we need to acknowledge that the Bible is an incredibly contextualised book. There is now an amazingly broad and complex range of realities that affect our 21st century lives, the majority of which we don't read directly about in Scripture. Yet just because the Bible doesn't mention electricity, air travel, computers, malaria, organ donation, Ebola, social media, IVF or genetically modified foods, it doesn't mean that these things are unimportant or that we shouldn't reflect on their use and impact in a biblical way. But this is rarely an easy task. Not only do we need to translate and interpret as we read the Bible for ourselves, we also need to listen to others and learn from their wisdom and experience.

With information and opinions flourishing everywhere, we must prayerfully discern who to trust as we seek to hear the voice of God for us today. Neglecting these practices makes it very difficult to properly understand what we are called to as 21st century disciples of Jesus of Nazareth. Additionally, it is worth keeping in mind that the Bible itself attests to the fact that it is not just church folk who can attain a profound grasp of the truth in certain situations. Indeed, if God can speak through pagan prostitutes as well as donkeys then it would pay to have very open ears, and an open – but critical – mind![89]

---

89 E.g. Rahab the prostitute in Joshua 2 and Balaam's donkey in Numbers 22.

Contemporary Prophets?

Looking no further than my own bookshelf it would seem that there has been no shortage of warnings issued about our lifestyles of material excess and the consequential environmental threats appearing on the horizon. In 1970 futurist Alvin Toffler's book *Future Shock* was doing the rounds on its way to becoming a best-seller. Many of his observations were quite astute, and his sense of urgency is clear:

> Our technological powers increase, but the side effects and potential hazards also escalate. We risk thermo-pollution of the oceans themselves, overheating them, destroying immeasurable quantities of marine life, perhaps even melting the polar icecaps. ... Through such disruptions of the natural ecology, we may literally, in the words of biologist Barry Commoner, be 'destroying this planet as a suitable place for human habitation.'[90]

> ... Here, then, is a pressing intellectual agenda for the social and physical sciences. We have taught ourselves to create and combine the most powerful of technologies. We have not taken pains to learn about their consequences. Today these consequences threaten to destroy us. We must learn, and learn fast.[91]

Toffler writes of the time being "late" – as if we were already running out of time – and that was more than forty years ago! He warned that "we simply can no longer afford to hurtle blindfolded towards super-industrialism"[92] and yet it would seem that we have done exactly that. A few years later in 1975 Bishop and theologian John V. Taylor conveyed similar sentiments with a matching urgency in his book *Enough is Enough*:

> The air, like the water which, as I have said before, man has always expected, like a nursemaid, to clear up his mess, has a strictly limited power of absorption. It cannot cope with the accumulation of toxic gases and particles in the quantities in which we now release them.[93]

---

90 Alvin Toffler, *Future Shock* (London: Pan Books, 1970), 388.
91 Toffler, *Future Shock*, 398.
92 Toffler, *Future Shock*, 402.
93 John V. Taylor, *Enough is Enough* (London, SCM Press, 1975), 32.

While Richard Foster is known for his teaching and influence on matters such as prayer and meditation[94], he made some radical statements back in 1981 in his book *Freedom of Simplicity*. His alarm at the ecological crisis humanity was facing is almost palpable:

> In spite of all the shortcomings of *The Limits to Growth* report of the Club of Rome, they did shout out to us one devastatingly undeniable truth: our 'growth mania' must stop or it will destroy us. We simply must understand that the wonderful resources of the earth are limited. We are now entering the age of scarcity. ... The message from all quarters is the same: our undisciplined consumption must end. If we continue to gobble up our resources without any regard to stewardship and to spew out our deadly wastes over land, sea and air, we may well be drawing down the final curtain upon ourselves. I need to add a reason for curbing our gluttonous consumption that Christians should consider very seriously. Overconsumption is a 'cancer eating away at our spiritual vitals.' It cuts the heart out of our compassion. It distances us from the great masses of broken bleeding humanity. It converts us into materialists. We become less able to ask the moral questions. [95]

Yet Foster, despite his alarm, was not without hope. Indeed he saw a crucial role for the Church as the world begins to comprehend and respond to the turmoil it has generated:

> Christians can make, I think, a unique contribution to this issue, because biblically and theologically we have a vital interest in both stewardship of the earth and economic justice for the poor. And because our allegiance to God is higher than to any nation-state, we have a commitment to global citizenship that can help us transcend the provincial claims of national interest.[96]

Interestingly, his suggestions for effective ways forward include a call for "international monitoring of the activities of global corporations" and that "strict international environmental standards must be set to govern the

---

94 Richard Foster's book *Celebration of Discipline* (1978) was named by Christianity Today as one of the Top 10 books of the 20th century.
95 Richard Foster, *Freedom of Simplicity* (London: Triangle, SPCK, 1981), 177.
96 Foster, *Freedom of Simplicity*, 178.

activities of multinational corporations."[97] I doubt many Christians would have expected such words from one of last century's spiritual giants! Such surprise demonstrates just how privatised and other-worldly much of our Christian thought and practice has become, and how politically polarised our thinking is between the 'right' and the 'left'. But are we not called to be salt and light in this world, and to stand up to injustice wherever we encounter it?[98]

A little over a decade later Tom Sine beckoned the church to respond proactively to the specific threat of climate change in his book *Wild Hope*. He proposed that we adopt a principal of caution, whereby we "take the threat seriously" – presuming there is truth in scientific claims and responding accordingly. While he acknowledged there was a chance that scientific theories about the global warming effect and its consequences might later be disproved, in his own mind the risks were so great that it was still worth taking precautions. This approach is surely prudent, particularly given his assessment that "Such a warming trend could increase the risk of forest fires, cause major droughts in farming regions, and even cause the oceans to rise."[99]

Interestingly, Tom Sine also wrote of a season of awakening about environmental issues during "Earth Decade", but it seems that this growing awareness in the late 1980's was unfortunately short-lived:

> We were all abruptly awakened from our environmental neglect by a series of events in the late eighties. In fact, in 1989 Time magazine devoted an entire issue to "Planet of the Year". The lead article read: "This year the earth spoke, like God, warning Noah of the deluge. Its message was loud and clear, and suddenly people began to listen, to ponder what portents the message held." ... Suddenly thoughtful Americans were rudely awakened to discover that our worst nightmares had come true. Our earth home is facing unprecedented catastrophes unless we act intelligently and decisively to change how we live on the earth. Never has there been such widespread and growing recognition by both leaders and those at the grassroots that we must clean up our act. The magic of the marketplace won't fix the environmental havoc we have wrought.[100]

---

97 Foster, *Freedom of Simplicity*, 181.
98 Matthew 5:13-16, Isaiah 58:6, Luke 11:42.
99 Tom Sine, *Wild Hope*, (Kent: Monarch Publications, 1991), 23.
100 Sine, *Wild Hope*, 18 & 19.

In *Wild Hope* Tom Sine goes on to list a range of environmental challenges, including the exploitation and degradation of forests, acid rain, global warming, ozone layer depletion, desertification, species extinction, chemical and toxic wastes, soil depletion, water pollution, burgeoning landfill waste, and the consumption of unnecessary consumer items.[101] While the science of climate change may not have had the strength of certainty and consensus that it does now, there was enough other evidence that we faced significant and urgent environmental threats that demanded our attention and response, and that climate change was worthy of inclusion among these. He concludes that,

> As inhabitants of this good earth, we have, I believe, ten to fifteen years at the outside to address these mounting environmental challenges. After that, I believe, we will be in danger of losing control of the processes of environmental degradation. The church in all its traditions must provide spiritual and moral leadership for this new environment movement. If we shirk in this responsibility, you can be sure that others with very different values will step into the breach. Therefore, those of us in the church, who believe we are called by our God to be earthkeepers, need to be in the forefront of this new movement for the protection and restoration of the created order.[102]

It is deeply saddening to read such clear and urgent pleadings from past decades and realise that we may well have failed to hear God speaking through his people. While some have demonstrated a general wariness about pagans, 'greenies' and environmentalists, as the quotations above indicate, there were respected voices from *within* the church who were proclaiming a similar message – and yet it seems we largely ignored them too. And, as was predicted, others have since stepped in to fill the gap. We can blame our fear, our pride, our hard and stubborn hearts, but in the end the important thing is to confess our sin: that we have failed to listen, to care and to act. We have been self-serving and greedy to the point that it has become idolatry. And the reality is that in doing so we have jeopardised so much, for so many. This seems to be the pattern with God's people – indeed there is nothing new under the sun! "The wisdom of the prudent is to give thought to their ways, but the folly of fools is deception. .... There is a way that seems right to a man, but in the end it leads to death." (Proverbs 14:8, 12).

101 Sine, *Wild Hope*, 30.
102 Sine, *Wild Hope*, 31.

In a similar vein, the American journalist Bill McKibben – now a prominent leader of global efforts to reduce carbon emissions, and also a committed Christian – offers this apt reflection:

> On the list of important mistakes we've made as a species, this one seems pretty high up. A single-minded focus on increasing wealth has driven the planet's ecological systems to the brink of failure, without making us happier. How did we screw up?[103]

Our many addictions and idolatries, our relative unfamiliarity with the horrors of war (at least for younger generations of Australians) and the deeply seductive power of the "Hollywood-ending" has, I believe, numbed our capacity to properly appreciate *dire threats*. Perhaps it is also human optimism, and certainly a good dose of pride, but somehow nearly all of us live with an overwhelming sense that "surely it won't happen to me". However, this does not reflect how things play out in the real world, particularly for those at the bottom of the social and economic ladders. The World Health Organisation estimated that by 2004 climate change was already causing 140,000 additional deaths per year,[104] and that was a decade ago!

The early impacts of climate change upon the world's poorest should come as no surprise to us. Prominent evangelicals such as Ron Sider spoke out about this decades ago, indicating that computer models of global warming trajectories gave a "tiny forewarning of a wrenching future", while noting that "in spite of this ghastly prospect, we keep adding carbon to the atmosphere at a dangerous rate".[105]

Avoidable tragedy and untimely death does not please God, and the statistics and projections of what lies ahead are genuinely alarming. And prominent economists such as Nicholas Stern have made it very clear that the costs of action on climate change are far, far less than the anticipated costs of inaction (indeed in 2009 Lord Stern revised his earlier estimates upward by 50%, suggesting that

---

103 Bill McKibben, *Deep Economy: The Wealth of Communities and the Durable Future* (New York: Times Books, 2007), 42.
104 World Health Organisation, *Climate Change and Health Factsheet No 266*, Reviewed November 2013, http://www.who.int/mediacentre/factsheets/fs266/en/, accessed 28 December 2013.
105 Ronald J. Sider, *Rich Christians in an Age of Hunger*, (Dallas: Word, 1990), 127 & 128.

the cost of inaction on climate change could equate to something in the vicinity of one third of global wealth)![106] In light of comments such as these, we are forced again to ask the question, "Why have we taken so very long to do so very little?"

In 2012 at the Doha Climate Summit Yeb Saño, representing the Philippines Climate Change Commission, urged delegates to

> ...let 2012 be remembered as the year the world found the courage to ... take responsibility for the future we want. I ask of all of us here, if not us, then who? If not now, then when? If not here, then where?"[107]

This was in the wake of Typhoon Bopha, which killed at least 500 people. Less than a year later Yeb Saño gave another impassioned speech, immediately following the even more devastating super Typhoon Haiyan.[108] In Warsaw he spelt it out even more clearly:

> To anyone who continues to deny the reality that is climate change, I dare you to get off your ivory tower and away from the comfort of your armchair. I dare you to go to the islands of the Pacific, the islands of the Caribbean and the islands of the Indian ocean and see the impacts of rising sea levels; to the mountainous regions of the Himalayas and the Andes to see communities confronting glacial floods, to the Arctic where communities grapple with the fast dwindling polar ice caps, to the large deltas of the Mekong, the Ganges, the Amazon, and the Nile where lives and livelihoods are drowned, to the hills of Central America that confronts similar monstrous hurricanes, to the vast savannas of Africa where climate change has likewise become a matter of life and death as food and water becomes scarce. Not to forget the massive hurricanes in the Gulf of Mexico

---

106   Michael McCarthy, "Lord Stern on global warming: It's even worse than I thought." *The Independent*, 13/03/2009. http://www.independent.co.uk/environment/climate-change/lord-stern-on-global-warming-its-even-worse-than-i-thought-1643957.html accessed 10 October 2014.

107   Amy Goodman, ""If Not Now, Then When": Filipino Negotiator Pleads for Climate Deal After Typhoon Kills 500", Democracy Now!, 7 December 2012.   http://www.democracynow.org/2012/12/7/if_not_now_then_when_filipino_ negotiator, accessed 26 November 2014.

108   Yeb Sano was the Philippines Chief Negotiator at the Warsaw 19th conference of parties (COP) to the United Nations Framework Convention on Climate Change, held in November 2013. Interestingly, during this speech Sano also committed to a voluntary fast that would last the duration of the conference (13 days): he was joined by many others at the conference, as well as thousands from around the world.

and the eastern seaboard of North America. And if that is not enough, you may want to pay a visit to the Philippines right now... What my country is going through as a result of this extreme climate event is madness. The climate crisis is madness. We can stop this madness.[109]

There are some clear reasons for our predicament and the profoundly inadequate global response so far. While it does not in any way absolve us of our responsibility, some very powerful forces have been at work in this period of history, contributing to an interesting and unfortunate unfolding of events. Understanding this complex interplay of money, power and politics does not provide *an excuse* for what has transpired, but rather a *broad explanation*. And this brief assessment is in no way a thorough treatment of the very serious issues involved. Rather, what follows presents some interesting starting points for those who want to dig deeper and reflect further, which for many of us is necessary training and preparation for the rather challenging road ahead.

---

109 RTCC, "It's time to stop this madness: Philippines plea at UN climate talks" *RTCC* (Responding to Climate Change) 11 November 2013. http://www.rtcc.org/2013/11/11/its-time-to-stop-this-madness-philippines-plea-at-un-climate-talks/#sthash.Y5UxS2Or.dpuf, accessed 20 Oct 2014.

3.3 Affluenza and "Gross" Domestic Product

Key Points:

We live in an age of rampant consumerism, akin to an epidemic disease.

Seeking to satisfy all our wants (rather than needs) is openly encouraged: in fact the "growth" of our modern market economy depends on it.

Our measure of national progress – GDP – is flawed and misleading.

With the able assistance of the fossil fuel industries, by failing to act on climate change we have jeopardised our very means of survival.

The idolatrous goal of 'economic growth at all costs' in no way reflects personal happiness or the achievement of human flourishing. It is far from the biblical model of love of neighbour, generosity, and sacrificial service.

Most Westerners would be aware that we live in a 'consumer society'. It is certainly not a foreign concept. Brilliant books and documentaries have demonstrated how addicted we are to our acquisition and consumption of 'stuff' – and generally people are not too surprised when you bring this reality to their attention.[110] They nod; they know. They realise that there is something not quite

---

110 When it comes to "stuff" one of my favourite resources is Annie Leonard's "Story of Stuff" – please take time watch her short on-line video if you haven't already: www.storyofstuff.com, accessed 26 October 2014.

right with our 21st century 'way of life' and yet they can't quite manage to escape its hold on them. Australian authors Clive Hamilton and Richard Denniss name this illness *affluenza*:

> Af-flu-en-za (n). 1. The bloated, sluggish and unfulfilled feeling that results from efforts to keep up with the Joneses. 2. An epidemic of stress, overwork, waste and indebtedness caused by dogged pursuit of the Australian dream. 3. An unsustainable addiction to economic growth.
>
> [...]
>
> Affluenza describes a condition in which we are confused about what it takes to live a worthwhile life. Part of this confusion is a failure to distinguish between what we want and what we need...[111]

Hamilton and Denniss highlight economic policy (based on a particular system of economic thought) as being one of the primary drivers behind this epidemic:

> Neoliberal economic policies have set out to promote higher consumption as the road to a better society. All the market-based reforms in the last two decades have been predicated on the belief that the best way to advance Australia's interests is to maximise the growth of income and consumption. No one has dared to criticise this. But the rapid expansion of consumption has imposed high costs, on consumers themselves, on society and on the natural environment... In addition to the rapid increase in consumer debt, higher levels of consumption are driving many Australians to work themselves sick. Yet our desire for various commodities (larger houses, sophisticated home appliances, expensive personal items, and so on) is continually recreated – an illness that entered a particularly virulent phase in the 1990s with the trend described as 'luxury fever'.[112]

Clive Hamilton's equally insightful book *Growth Fetish*, describes how economic growth has become the new "sacred" aspect of society:

> The development mentality is the daily manifestation of growth fetishism. Every day governments and local authorities approve

---

111 Clive Hamilton & Richard Denniss, *Affluenza* (Crows Nest: Allen & Unwin, 2005), 7.

112 Hamilton & Denniss, *Affluenza*, 8.

housing developments, shopping malls and roadway projects that despoil the remaining natural areas. The momentum of development seems to mesmerise decision makers. In a political environment where growth is regarded as a sacred duty of elected leaders, blocking a new shopping mall or a housing development is akin to taking a stance against progress itself, a challenge to one of history's immutable laws that few mayors or planning ministers are willing to embrace.

[…]

Protecting economic growth, and the system of private property on which it is based, has become one of our most powerful impulses, one that neoliberal leaders trade on relentlessly. It is a religious urge.[113]

Thus, the present movement toward growing one's own food would not be considered as being in the "national interest". Saving seeds and learning precious life-skills that enable households to produce food for their own tables means we purchase less from the formal economy. If adopted widely, this would lead to a decline in GDP (Gross Domestic Product) which would be – using our normal measures of progress – a *bad thing*. In contrast, eating highly processed foods (including the hundreds of "food miles" clocked up in the various phases of production, processing, packing and delivery to mega-markets everywhere) would be a *good thing*: higher GDP generally equals more jobs which in turn facilitates increased levels of consumption. And highly processed foods generally means less health, which is also great for the economy: the more we visit doctors, take supplements and medications, undergo medical treatments, and purchase merchandise promoted as being essential to the latest dieting or fitness fad, the more our economy thrives! Surely this example alone is proof that our story about economic growth and progress needs some serious re-visioning!?

Interestingly, Hamilton goes on to explore the connection between the religious fervour of the "purest defenders of capitalism" and the Genesis story whereby creation is given to mankind, suggesting that for those who pursue this neo-

113 Clive Hamilton, *Growth Fetish* (Crows Nest: Allen & Unwin, 2003), 184–185.

capitalist agenda "environmentalists have displaced communists as the devil incarnate". While most Christians would not agree with this proposition, it is important for the Church to hear such critiques so that we can make an intelligent, informed response.[114] He goes on:

> Although they are not generally conscious of it, the radical Right are the intellectual keepers of the Old Testament canon according to which God created Nature for the benefit of man. Calls by environmentalists to respect the integrity of the natural world, and on this basis to stop certain commercial activities, are met with incomprehension and rage. The right to exploit the land is, after all, God-given. This toxic mix of divine endorsement and private property rights is most apparent in the ideology and political activity of the mining and agricultural industries.[115]

Again, while Christians would generally not support claims that the Bible presents a natural world that exists purely for the benefit of humankind, there is certainly wisdom in examining whether we may have naively bought into attitudes promoted by the prevailing neo-capitalist ideology.

There is warrant in calling our attention to the presumptuous claims and activities of the mining and agricultural industries. These two industries are crucial to our present 'way of life' and in recent decades they have amassed amazing corporate and political clout. At present we are, unfortunately, at their mercy in a number of ways. Firstly, our reliance upon fossil fuels makes us beholden to our mining industry: as things stand they sustain our habits, our comforts, and our travelling convenience – still at such "affordable" prices. Secondly, very few of us are able to practice even the basics of self-sufficient living: we are unfamiliar with what it really takes to grow our own food, and accordingly are now largely depend upon a food production system run by multi-national agri-businesses! Farmers' markets are lovely, and are certainly a step in the right direction, but can you imagine the chaos if our mega-supermarkets closed and we suddenly had to fend for ourselves? We are stuck living within a complex system that has its own interests (profits) in mind rather than the welfare of people and their environment.

---

114 Lynn White's thesis written in the 1960's is a similar example, which we explore further in this section.
115 Hamtilon, *Growth Fetish*, 185-186.

Along these lines, a number of economists have been working on numerical alternatives to GDP that include a broad range of factors that influence our wellbeing and provide a more accurate measure of national progress. One of these is the Genuine Progress Indicator (GPI) or the Index of Sustainable Economic Welfare. While it is still consumption-based and imperfect, such a model points us in a more helpful direction and simultaneously sheds light on just how unhelpful our obsession with GDP really is. Some of the additional factors that are included in measures such as the GPI are the value "work" that is conducted within homes and the general community. Such work falls outside the boundaries of the formal market as there is no price tag attached. This includes raising children, caring for the sick and elderly, housework, meal preparation, and community service. Without commercial value they are excluded from normal measures – as are environmental "costs" such as logging, greenhouse gas emissions, and ozone depletion (which are all incorporated into the GPI, but not the GDP).

Growth Fetish discusses some of these measures in more detail.[116] Interestingly, while graphs of GDP show a reasonably consistent upward trend for the UK and USA for the latter half of last century, the GPI shows a peak in the late 1970's and a reasonably consistent decline in the years since. It is striking to see a graphical illustration of the fact that we are – quite possibly – no longer making the progress we believe we are working so hard to achieve. But in terms of our national leaders, few seem to notice or care.

While I am at times more cynical than Christian author Tom Sine, who in 1999 suggested that the economists and politicians who advocate most strongly for "a new global economic order" do so because they genuinely "want what is best for our planetary community", I would certainly agree that they do what they do because they "have been schooled in a worldview that defines what is best largely in terms of economic growth and economic efficiency."[117] Even those who seek their own financial gain do so because they believe it will somehow be of immense benefit to them. Beyond a reasonably simple annual income however, researchers have found that there is no obvious correlation between material wealth and personal happiness. Or as Bill McKibben puts it: "money

---

116 Hamtilon, *Growth Fetish*, 54ff.

117 Tom Sine, *Mustard Seed versus McWorld: Reinventing Life and Mission for a new Millennium* (London: Monarch Books, 1999), 64.

consistently buys happiness right up to about US $10,000 per capita income"[118] – then the strong correlation falls away.

While the resources that money can buy certainly have value, particularly for those who are in dire need, we are now fully accustomed to meeting much, much more than our basic needs. Tom Sine examines this issue further, and while some might claim that it is this expansionist market model that will bring billions of people out of poverty, he makes it very clear that the underlying aspirations and values reflect those of *modern culture*, and that they are in many ways in direct conflict with those at the foundation of a *biblical faith*.[119]

Time and time again the teaching of Jesus as recorded in the Gospels presents a message of seeking first the Kingdom of God – characterised by the pursuit of justice and rightness and of loving each other – in contrast to a greedy, futile and fraught pursuit of stuff. The ancient teachings of the Son of God and his new Kingdom provide a profoundly relevant wisdom for us today.

A crucial point that is made often within environmental circles is that it is highly illogical, irrational and fundamentally unsustainable to pursue an agenda of *infinite growth on a finite planet*. There is only so much plundering and polluting that can be done before planet earth caves under the strain. The planet itself would survive, but many of its precious species would not – perhaps even a significant proportion of humankind. However, our collective attachment to growth is so strong that we find it hard to think seriously about the reality and the ramifications of such claims.[120] Clive Hamilton makes a very clear call about this cause and effect relationship:

> There has been one, and only one, reason for the reluctance of the rich countries of the world to reduce their emissions and so help to stave off environmental catastrophe – the perceived impact of reducing emissions on the rate of economic growth and especially the growth of a handful of powerful industries. *This has been enough to jeopardise the future of the world.*[121] (emphasis mine)

---

118 McKibben, *Deep Economy*, 41.
119 Tom Sine, *Mustard Seed versus McWorld*, 67.
120 Clive Hamilton, *Growth Fetish*, 175.
121 Clive Hamilton, *Growth Fetish*, 182.

While this may appear both extreme and simplistic, there is mounting evidence to support such views. As Bill McKibben articulates well in his recent book *Oil & Honey*, the reasonably straight-forward goal of corporations is to make profit, with little regard for much else: their relentless and simplistic obsession with profit maximisation "will combine with their wealth to overwhelm reason, science, love."[122]

It is now common knowledge that the same PR firms who were paid to (mis) lead the public to believe in the apparent safety of tobacco have more recently been paid to generate doubt about the certainty of the science regarding climate change. They weren't given the task of refuting the science. No, that strategy would backfire on them far too quickly, as the fossil fuel industries knew that the scientists had actual science on their side. Rather, these firms were paid simply to create uncertainty, which would raise sufficient suspicion to stifle progress on effective action.[123] And this is exactly what they have achieved.

Before we move on to look briefly at the intertwined role of marketing and the media, it is worth stepping back even further to consider the extent to which we have genuinely 'progressed' as a society. While many white settlers worked hard to 'civilize' Australia's indigenous peoples, many would now agree that we have neglected to learn from them to our own detriment.

Some older cultures are certainly rich with wisdom concerning sustainability. More than four decades ago H.C.Coombs, a well-known Australian economist and public servant, warned that our "hectic rate of change" and "persistent pressure to the consumption of goods" were significantly eroding our identity and sense of purpose. In a lecture given in 1971 he compared the very evident social fragmentation of modern, Australian culture with the much more sustainable models of Australia's first people. As quoted in the book *The Return of Scarcity*, Coombs states that:

> The Aboriginal people of Australia lived, and some still live, in a society of extreme material simplicity. They found both security and challenge in winning a reluctant livelihood from an inhospitable land. They found time for an artistic and ceremonial life of a richness we can but envy.

---

122 Bill McKibben, *Oil & Honey: The Education of an Unlikely Activist* (Collingwood: Black Inc., 2013), 104.

123 Merchants of Doubt by Naomi Oreskes and Erik Conway is a key 'text' on this issue. More information can be found at "Merchants of Doubt" (http://www.merchantsofdoubt.org/) as well as 'Climate Cover-Up: The Crusade to Deny Global Warming' by James Hoggan with Richard Littlemore and the "DeSmogBlog" (http://www.desmogblog.com/).

> In their oral traditions they built and preserved a fabric of myths expressed in story, song and ritual which expressed their intimate relationship with the land and its creatures. From Captain Cook onwards, those who have known them have wondered at their freedom from the tyranny of things and at the steadfastness of their moral and spiritual values. [124]

While we must avoid the tendency to romanticise indigenous peoples and cultures, there is yet an important truth in such a statement. It is increasingly apparent that individualistic consumerism has eroded our ability to live rich, connected and truly meaningful lives. In a similar vein, indigenous leader Ivan-Tiwu Copely (a Peramangk/Kaurna man who was a recipient of the Order of Australia Medal in 2012) wrote recently,

> As Aboriginal People we have a close living relationship with the Land and have understood this through learning over thousands of years. We have looked after this Land, as a child would look after its Mother and pay it great respect. And in return it has looked after us like children and nourished us for thousands of years... *as our Land dies, so do we.* [125]

While we have certainly made tremendous advances in all kinds of areas (think of modern medicine and communications technologies as a start) we can too easily assume that our way of life is the *only* way, or that it unquestionably represents the *best* way. Indeed our busy private and professional lives, our demanding family schedules and our media-saturated homes generally prohibit us from creatively exploring any alternatives. Instead of providing us with abundant life and leisure, the bountiful booty of the 'free' market instead makes us feel stifled, stuck and sometimes even sick.

---

124 H.C. Coombs, *The Return of Scarcity* (Cambridge: Cambridge University Press, 1990), 59. Interestingly, Coombs was the Director of Rationing in 1942 during WW2, which was undoubtedly an invaluable experience in terms of understanding the difference between wants and needs!

125 From written (email) correspondence, 23 October 2013 (emphasis mine).

3.4 "Biggering and biggering": market and media

Key Points:

Our economy relies heavily upon the advertising industry to generate new "needs" which it then satisfies. *Contentment is the enemy of the market.*[126]

Advertising revenue is now what funds a significant proportion of our news and entertainment. *Critical thinking is an enemy of the market.*

Corporations are so intent on growing that they have demonstrated unscrupulous behaviour in exploiting new markets for their goods & services. Disasters and conflict are increasingly profitable: *Peace is an enemy of the market.*

And to protect their massive commercial interests, the fossil fuel industry has funded an extensive misinformation campaign in order to ensure that the science of climate change is not acted upon. *Truth is an enemy of the market.*

In 1971 Dr Seuss wrote a children's book called *The Lorax*, which offered a timely word of prophetic insight and warning to children and adults alike. In this story, the "thneed" is a garment that nobody needs, made from the beautiful "truffular tree". Due to the high demand for thneeds, this unique plant species faces extinction, the air and water becomes polluted and the land is soon uninhabitable. The narrator of the story, who was once the entrepreneurial

---
126 Note that here I use the term *'market'* as short-hand for "the contemporary system of growth-obsessed market-based consumerism".

businessman responsible for the production of thneeds, relates to his audience the story of his blind pursuit of money:

> I meant no harm. I most truly did not.
>
> But I had to grow bigger. So bigger I got.
>
> I biggered my factory. I biggered my roads.
>
> I biggered my wagons. I biggered the loads
>
> of the Thneeds I shipped out. I was shipping them forth
>
> to the South to the East! To the West to the North!
>
> I went right on biggering... selling more Thneeds.
>
> And I biggered my money, which everyone needs.[127]

In 1958, economist Kenneth Galbraith wrote about what he termed "The Dependence Effect". In response to claims that our capitalist economic system provided the most effective and efficient ways to meet people's urgent needs, he warned that there was a fundamental flaw in this argument. When the very system of production includes the creation of artificial "wants" that it then seeks to satisfy, the model can no longer be deemed efficient or effective. Put more simply: "One cannot defend production as satisfying wants if that production creates the wants".[128] People have obvious needs for food, water, shelter, clothing, and health care, as well as the need for trusting relationships, meaningful work and freedom from violence.

It is right for societies to explore the best ways to meet these needs, and economics has an important role to play in these deliberations. The capitalist model empowers individual or corporate producers to find profitable ways to meet these needs for us, rather than this being a task of the state or the community. The flaw that Galbraith identified is that these producers have increasingly adopted a model whereby their own marketing campaigns create artificial needs (i.e. wants) which they then very happily and ably satisfy.

---

127 Dr. Seuss and A.S. Geisel, *The Lorax*, (New York: Random House, 1971), 38.

128 J.K. Galbraith, *The Affluent Society* (Mitcham: Penguin, 1958), 131.

The entrepreneur in Dr Seuss' story saw an opportunity to make money and seized it, creating a need for thneeds in society. A recent example of this phenomena in the real world is the "onesie" trend: the over-sized, one-piece garment that suddenly became all the rage for kids, youth and even adults! People's wardrobes were already over-full (and their credit cards probably maxed-out too) yet another fashion 'trend' made people feel either jealous or inadequate without possessing this latest, greatest thing.

Whose life has been genuinely and significantly improved by a onesie? More resources used to satiate the endless consumption of fad products inevitably leads to more waste once the fad passes, as the neglected onesies gradually make their way to op shops and then eventually into landfill. Meanwhile people in many places around the world lack access to drinking water and basic sanitation. Added to this is the growing tendency for designers to incorporate planned obsolescence into new products (whereby products have an artificially limited useful life, requiring early or frequent replacement) and we have a recipe for environmental disaster! What have we become?

In its earliest forms and up until 1890, advertising was "an essentially serious and rational enterprise whose purpose was to convey information and make claims in propositional form primarily about providing information."[129] It sought to appeal to people's understanding, not their passions. With the introduction of illustrations, photographs, slogans and jingles at the turn of the century this all began to change, and "advertisers no longer assumed rationality on the part of their potential customers."[130]

One century later the advertising industry is a massive force to be reckoned with. Direct advertising as well as corporate sponsorship means that ads are a part of nearly every event and media source you could think of, including sporting events, bus-stops, TV, magazines, newspapers, websites and social media. Even schools and children's sporting clubs are coming to rely more and more heavily

129 Neil Postman, *Amusing Ourselves to Death* (2006 edition with introduction by Andrew Postman, Camberwell: Penguin), 59-60.
130 Postman, *Amusing Ourselves to Death*, 60.

upon various forms of sponsorship and the branding that generally accompanies it. And the commercial and political ties with media are stronger than ever. The purchase price for a commercial newspaper does not begin to cover the cost of journalists, editors, content, printing and distribution. It is advertising revenue that makes it all profitable and possible.

It is logical that the flow-on effect is that our "news" is consistently tainted by an array of vested interests.

In speaking with a local journalist (who was bemoaning the lack of staff available to cover a local climate change rally) I learnt that a few decades ago there were four journalists for every PR professional. Now the trend has reversed: there are four PR professionals for every journalist! These PR pros are paid and resourced – generally by large corporations with big budgets – to ensure that the right 'news' makes it into the media. Often this news is that which serves particular commercial interests – to the point that it can now be hard to discern the difference between genuine news and an infomercial.

It certainly doesn't mean that there is no genuine news anymore. The very few investigative journalists left are, however, run off their feet. They struggle to follow up useful leads and stories of genuine social interest, and are undoubtedly censored at various points where their news story might offend the sensitivities of important advertising clients or large, powerful lobby groups.

So it does seem that we will get what we pay for: if you pay nothing for your news you may find that it comes served with a very large 'side' of advertisements, infomercials, and special offers. And of course the whole meal might be strongly flavoured with political bias, depending on who owns the media! So it is no surprise that many non-commercial media (e.g. government-funded sources such as the ABC/SBS, and university funded sources such as The Conversation) often do a considerably better job reporting on climate science than the commercial media.

Clive Hamilton draws attention to the significant disconnection between public opinion and political action on climate change, and points the finger directly at the media and their "long-term failure to give climate change the attention it deserved."[131] He gives a poignant example of just how poorly the issue of climate change has been handled, even to the point of it becoming a marketing gimmick:

> The Sun-Herald's travel section carried a cover story identifying the ten most spectacular natural sights that may soon disappear as a result of global warming and urging readers to see them while they could… Whereas others see climate change as a serious threat to tourism, the industry has embraced it as a marketing opportunity.[132]

And there is a vicious cycle at work when it comes to media, marketing and entertainment (which could readily include tourism). Entertainment sells advertising, which in turn generates revenue for advertisers and customers for corporations. Furthermore, our entertainment-saturated society tends to lull us into a state of passivity and indifference. As a consequence we increasingly struggle to think rigorously and critically about matters of real substance. Neil Postman wrote brilliantly and lucidly about this concerning trend in his brilliant book *Amusing ourselves to Death*. Back in 1975 he warned that the ascendance of the TV as a primary way of learning about the world meant that much public discourse had become "dangerous nonsense" with people becoming "sillier by the minute".[133] We are forgetting how to read, and consequently how to think critically. And this suits corporations just fine: so long as we don't spend so much time on our couches that we forget to consume (less of a risk now with the rise of TV and internet shopping and express home delivery)!

The onesie is a reasonably benign example of the highly aggressive and competitive push by increasingly globalised corporations to find new products and new markets. "Biggering and biggering" is the aim of the game, and merely more of the same old stuff won't facilitate the levels of growth upon which the free-market system thrives.

---

131 Hamilton, *Scorcher*, 161.
132 Hamilton, *Scorcher*, 164.
133 Neil Postman, *Amusing Ourselves to Death*, 16, 24.

Naomi Klein makes a significant and disturbing connection between climate change and what she calls the "disaster capitalism complex". In summary, disaster capitalism is characterised by "orchestrated raids on the public sphere in the wake of catastrophic events, combined with the treatment of disasters as exciting market opportunities".[134] While the rumours she heard while in Sri Lanka of the US being behind the tsunami are very unlikely to contain an ounce of truth, it does demonstrate that many people are becoming incredibly suspicious of the large-scale, post-disaster economic interference that tends to end up making the rich even richer. The exploitation that follows disasters is now so obvious that there is growing suspicion on the ground that these events are being deliberately orchestrated by the rich and powerful, who very often end up walking away with spectacular profits at the end of the day. As Klein explains:

> An economic system that requires constant growth, while bucking almost all serious attempts at environmental regulation, generates a steady stream of disasters all on its own, whether military, ecological or financial... Our common addiction to dirty, non-renewable energy sources keeps other kinds of emergencies coming: natural disasters (up 430% since 1975) and wars waged for control over scarce resources (not just Iraq and Afghanistan but lower intensity conflicts such as those that rage in Nigeria, Colombia and Sudan), which in turn create terrorist blowback... Given the boiling temperatures, both climactic and political, future disasters need not be cooked up in dark conspiracies. All indications are that simply by staying on the current course, they will keep coming with ever more ferocious intensity. Disaster generation can therefore be left to the market's invisible hand.[135]

More concerning still is the connection between large oil companies (who have poured tremendous amounts of money into promoting climate change denial), the weapons industry, homeland security contractors, and more recently the media industry. This new kind of corporate synergy is emerging rapidly in the US but spreading its influence around the globe:

---
134 Naomi Klein, *The Shock Doctrine: The Rise of Disaster Capitalism* (Camberwell: Penguin, 2007), 6.
135 Klein, *The Shock Doctrine*, p. 426-427.

It certainly makes sound business sense. The more panicked our societies become, convinced that there are terrorists lurking in every mosque, the higher the news ratings soar, the more biometric ID's and liquid-explosive-detection devices the complex sells, and the more high-tech fences it builds... The only prospect that threatens the booming disaster economy on which so much wealth depends – from weapons to oil to engineering to surveillance to patented drugs – is the *possibility of achieving some measure of climate stability and geopolitical peace.*[136] (emphasis mine)

It is abundantly clear that much needs to change, and soon. While some might still nursing long-held concerns of an impending financial collapse related to the price on carbon (we were fine, by the way) or be worried about a massive slowing of the economy due to lower rates of personal consumption, the writing has been on the wall for more than 40 years: *the path we are on is fundamentally unsustainable.*

Corporations and the fossil fuel industry have had decades to plan accordingly and to map out a more sane and sustainable way forward, but they have instead chosen to hedge their bets in all the worst ways. They have invested more and more in long-term projects that rely upon ongoing extraction and combustion of fossil fuels, using methods that are simultaneously less efficient and more hazardous (including coal seam gas/fracking and shale oil). They have simultaneously funded think-tanks that promote climate change denial.

On a trajectory of four degrees of warming (which we are likely to exceed, on current trends) human-induced climate change now threatens the survival of the majority of the species on the planet. Is our fight going to be for our growth-addicted economy and the very few for whom it is profoundly profitable, or the very real world – the one that God made and called good – that at present sustains all life on earth?

---

136 Klein, *The Shock Doctrine*, 427-428.

Comment by Byron Smith (an Anglican Minister who is currently completing his PhD in theological ethics, and friend to Mick and Claire): "There is now no non-radical option available. Either we have radical economic and political change in order to avoid catastrophic climate change, or we walk into radical climate change and the radical social, economic, political and ecological consequences it entails. When we talk of something being unsustainable, what we ultimately mean is that *it will not be sustained*. Something must change: will we choose changes to our human systems that have other desirable features, or have involuntary ones imposed on us as the planet changes? Changing the laws of politics and economics might seem impossible, but it is easier than changing the laws of physics and chemistry."

And before anyone thinks to call this kind of thinking "radical", consider these words from Bill McKibben:

> "The radicals are the people who are fundamentally altering the composition of the atmosphere. That is most radical thing that people have ever done."[137]

---

137 Bill McKibben, in a speech given at Power Shift 2011. http://grist.org/climate-change/2011-04-18-bill-mckibbens-must-watch-speech-at-power-shift/, accessed 20 June 2014.

3.5 Polluted Politics

Key Points:

In Australia, climate change has become an increasingly polarised and politicised issue.

Not only have we been duped by our political leaders, we have missed out on tremendous opportunities to make the most of our natural assets and to invest in Australia's long-term prosperity.

In seeking to protect Australia's short-term national interests and pander to our very powerful resources sector, our political leaders undermined the first genuine, global attempt to tackle climate change (the Kyoto Protocol, which was adopted in 1997).

The claim that we are small and significant in relation to global emissions has been used as a smoke screen to hide the fact that we are among the world's most consumptive and environmentally unfriendly people.

The neo-liberal economic experiment and the vast power of corporations and media are now very much a global enterprise. I am convinced that these factors have conspired to inhibit climate action in a number of ways. But when it comes to politics, there are a few countries where democratically elected governments are failing monumentally in terms of leadership on climate change. Australia would be close to the top of this list, if not at the very top. It would seem that perhaps the world has been waiting for the United States (the nation that has historically topped the 'big emitters' list, based on both absolute and per-capita emissions) to make really solid progress in this area, despite the State of California and a handful of other US States powering ahead regardless.[138]

However the significant emissions reduction agreement between the US and China (reached in the lead up to the G20 held in Brisbane in November 2014) is a clear indication that the policies of nations with big emissions are certainly shifting in the right direction. Few Australians, however, seem to be aware of just how unhelpful our national approach to this issue has been – to the point that we would be the laughing-stock of the world… were this issue not quite so serious! What follows is a snapshot of a particularly disappointing period of Australia's political history on climate change policy.

At the World Economic Forum held in 2000, global warming was declared "the 21st century's greatest challenge". One would think that this would have provided a clear call to both industry and government to not simply sweep this issue aside. Yet in his book *The Third Degree*, Murray Hogarth makes the following observation: "Many industries and companies were complicit through ignorance, apathy and inaction, letting years slip by without engaging on the climate challenge, taking their cue from Canberra."[139] Hogarth's book was written in 2007 in the context of a perceived window of opportunity, where there was a growing optimism about action on climate change. During the previous year many Australians had "Walked Against Warming" in rallies held around Australia. *An Inconvenient Truth*, starring Al Gore, released in 2006, had begun to put the issue into the mainstream. There was a sense of momentum and hope.

But all of that unravelled incredibly quickly. In late 2009 Australia's long drought finally broke, Abbott defeated Turnbull by one vote to become Liberal Party leader (with the primary focus of the leadership spill being Turnbull's support of an Emissions Trading Scheme) and then shortly after this the Copenhagen summit failed to achieve any real progress toward binding international

---

138 The internet has no end of articles and resources about California's efforts to "Go Green". The best place to start is probably the State's own website: http://www.green.ca.gov/ - Vermont and Washington are also making commendable efforts, along with Hawaii.

139 Murray Hogarth, *The Third Degree* (North Melbourne: Pluto Press, 2007), 8.

emission reduction targets. While stating upon his election to the position of party leader that he believed climate change to be real and that "man does make a contribution"[140], Prime Minister Tony Abbott would later repeal the price on carbon, effectively setting the clock back decades.

Meanwhile developed economies were still reeling from the Global Financial Crisis of 2007/2008: action on climate change was the least of their worries! Simultaneously, the amazing growth of the Chinese economy saw the price of Aussie coal soar. Climbing electricity prices and the politicisation of the price on carbon during the era of the Rudd/Gillard governments combined to well and truly close this particular window of opportunity for effective climate action. So while Tony Abbott spent his time in opposition bemoaning the cost of the "big, new tax on everything" and how it would cripple our industries and undermine our way of life (which, in hindsight, it clearly didn't), we heard next to nothing about the cost of climate impacts upon agriculture, water security, health and infrastructure – which are actually vitally important facets of *our way of life*. As Age columnist Jonathan Holmes laments, "Labor doesn't talk about that stuff either, these days. We heard it all six years ago. Alarmism. Warmism. We've moved on."[141]

Clive Hamilton's book *Scorcher – The Dirty Politics of Climate Change* provides a very eye-opening assessment of the Coalition Government's climate policy during the Howard era (1996-2007), including some of the very strong links between the fossil fuel industries, the 'greenhouse mafia' that was tasked with ensuring inaction on climate change, and our own government. The tragedy of this is summarised well in the introduction:

> When it comes to cutting greenhouse gases, rather than being at a unique *disadvantage* compared with other industrialised and developing countries, as the Government claims, Australia is particularly well placed

---

140 Ari Sharp, "Abbott wins Liberal leadership – by one vote", *The Sydney Morning Herald*, 1 December 2009. http://www.smh.com.au/national/abbott-wins-liberal-leadership--by-one-vote-20091201-k1va.html, accessed 10 October 2014.

141 Jonathan Holmes, "From great moral challenge to indifference", *Sydney Morning Herald*, 4 September 2013. http://www.smh.com.au/federal-politics/federal-election-2013/from-great-moral-challenge-to-indifference-20130903-2t355.html, accessed 30 September 2014.

for the longer term global energy revolution. Exploiting this advantage means seizing the opportunity at an early date to build on Australian expertise already acquired in areas such as energy efficiency, and solar technology, and, more generally, moving early to reduce the long-term costs of conversion to non-fossil-fuel sources. However, as this book will document, the Australian Government's position on climate change has taken us in the opposite direction, emphasising short-term economic goals and entrenching fossil fuels in the nation's energy economy. It has undermined research and commercial initiatives leading to a low-carbon future. Going back as far as Federation, it is hard to think of a political and policy failure that has been so damaging to the long-term national interest and so reckless in its disregard for the welfare of future generations of Australians.[142]

There are some glaring examples of the conflict of interest and intentional misinformation during this era of government. For example, Queensland Senator Warwick Parer was the Minister for Energy and Resources in the lead up to the Kyoto negotiations of 1997, despite having a background of significant involvement in the coal industry (including a $ 2million commercial interest in a company that owned three coal mines). Parer decided to abolish the Energy Research Development Corporation, which was a body that had been tasked with developing energy technology that could replace fossil fuels. Hamilton reflects,

> The fact that Howard appointed as the minister for resources and energy a man who rejected greenhouse science, defended the interests of the coal industry at every opportunity and had a large personal financial stake in coal mining was symbolic of his approach to climate change.[143]

ABARE (Australian Bureau of Agricultural and Resource Economics) was the body tasked with producing models to facilitate the development of climate change policy, however in May 1997 extremely damaging criticism of their work emerged when it was revealed to the Senate that:

---

142 Clive Hamilton, Scorcher: *The Dirty Politics of Climate Change* (Melbourne: Black Inc. Agenda, 2007), 19.
143 Hamilton, *Scorcher*, 59-60.

most of the funding for the research had been received from the fossil-fuel industry, including the Australian Coal Association, the Australian Aluminium Council, BHP, CRA, the Business Council of Australia, the Electricity Supply Association of Australia, Exxon, Mobil and Texaco. These organisations paid $50,000 annually for a seat on the steering committee.[144]

Yet when the Australian Conservation Foundation applied for membership of this committee, requesting that this hefty fee be waived, they were refused. And in June 1999 the Prime Minister's Science, Engineering and Innovation Council urged the government to get proactive on climate change policy, issuing a report that describes Kyoto "as a watershed in the global greenhouse debate" and arguing "that it was a powerful instrument of change that would be ignored at great cost."

Instead of listening, the Government changed the membership of the Council to ensure it would receive more appropriate advice in the future (including the appointment of Robin Batterham as chief scientist and chair, while he continued as Chief Technologist for Rio Tinto).[145]

The public was serially misinformed with regard to the cost of action on climate change, including statistical tricks employed by ABARE. Two examples include:

> The Government claimed that Australia faced ruinous costs to address greenhouse gas emissions, while failing to point out that "If Australia reduced its emissions, according to the estimates, the doubling of per capita incomes would have to wait until around 1 March 2025"– which happened to represent "a delay of a mere two months"![146]

> On the eve of the Kyoto conference in 1997, Howard announced a *major climate policy initiative* which included expenditure of $180 million. While it appeared impressive, this funding would be allocated over a five year period which meant that annually it would equate to the price of a single

---

144 Hamilton, *Scorcher*, 62-63.
145 Hamilton, *Scorcher*, 124.
146 Hamilton, *Scorcher*, 61.

bus ticket, per Australian per year! Many rightly felt that this response was "entirely incommensurate with the seriousness of the problem."[147]

A large portion of *Scorcher* is devoted to unpacking Australia's involvement in the development of the Kyoto Protocol, which is very much a story in its own right. This was a crucial time in terms of being the first global attempt to make progress toward meaningful solutions. Australia's part in it can only be described as shameful. Here is a summary:

~ *Australia pushed for generous concessions.* Our nation's Government adopted a bullying tactic to threaten the established consensus: Australian delegates and leaders had spent months prior to the conference floating talk of withdrawing support. These mean and self-serving tactics were labelled "wrong and immoral" by other nations – but they worked! It had been suggested the emission reduction "burdens" should be proportional based on per capita emissions and relative national wealth (the "polluter pays" and "ability to pay" principles – see over page for definitions). Following such principles would mean that Australia should have more *stringent* targets imposed than nearly every other country, yet they managed to walk away with targets that were in fact far more lenient![148]

~ *The Government repeatedly ignored calls to support the Kyoto Protocol.* As well as changing the membership of the Innovation Council (as already discussed), they also ignored a statement signed by more than 250 of Australia's academic economists which called for an immediate support for the Protocol. This move "directly challenged the Government's rationale for refusing to ratify Kyoto on economic grounds". Instead of supporting the Howard government's line that strong action on climate change would threaten the national interest, these economists saw that climate change presented such serious risks at environmental, economic and social levels that preventative steps were certainly warranted.[149]

---

147 Hamilton, *Scorcher*, 69.
148 Hamilton, *Scorcher*, 75-79.
149 Hamilton, *Scorcher*, 123.

~ *The Government repudiated the Protocol and refused to sign* (yet somehow Senator Ian Campbell managed to assert that "no one has shown more support for the Kyoto Protocol than Australia")[150]

~ *The Australian Government then pushed to be included on the Ad Hoc Working Group on Future Commitments*, which had had been tasked with negotiating the structure of the Kyoto Protocol during the second commitment period (from 2013). So, despite having won amazingly generous concessions for the first commitment period – so generous, in fact, that it didn't actually have to achieve anything at all – Australia refused to ratify the Protocol, and then cried foul when it had no capacity to influence further negotiations.[151]

~ *A hypothesis emerged that perhaps the Howard Government had been actively trying to destroy the Protocol in order to preserve its coal markets*: Due to its significant coal export industry, Australia stood to lose a lot from efforts by other countries to reduce their own emissions. This hypothesis helps to explain a paradox that has bothered many commentators, including the European negotiators. If Australia had repudiated the Kyoto Protocol, why was it so desperate to continue to participate in Kyoto negotiations? Why did it want to shape a treaty that it had rejected outright? A plausible answer was that as long as the Howard Government had a seat at the table, it could continue to spoil things and make overall progress more difficult.[152]

"The polluter pays principle (PPP) identifies the parties who caused the pollution and apportions responsibility for paying the costs of dealing with climate change among those parties. Arguably, the PPP is the most intuitive way of thinking about the ethics of climate change. It is based on the widely shared idea that those who cause harm to others should be morally responsible for remedying that harm. As such, the PPP has the ability to provide the appropriate incentive to prevent polluting by directly linking moral responsibility, and the resulting accountability, to the kinds of actions that should be discouraged."[153]

150 Hamilton, *Scorcher*, 171.
151 Hamilton, *Scorcher*, 177.
152 Hamilton, *Scorcher*, 223.
153 Dan Weihers, David Eng and Ramon Das, "Sharing the responsibility of dealing with climate change: Interpreting the principle of common but differentiated responsibilities." *Public Policy: Why ethics matters*, ed. Jonathan Boston et al, (Canberra: ANU E Press, 2010). http://press.anu.edu.au/apps/bookworm/view/Public+Policy%3A+Why+ethics+matters/5251/upfront.xhtml, accessed 10 October 2014.

> "The ability to pay principle (APP) regards states' per capita production capacity (or some other measure of welfare) as the only moral consideration in sharing the responsibilities of remedying the adverse effects of climate change. The APP requires that all and only those who can afford to pay for mitigating and adapting to climate change should pay and they should pay in proportion to their ability to pay."[154]

Clive Hamilton offers this scathing assessment at the conclusion of his book:

> It is painful to be a citizen of a nation that could behave in such an immoral way, but the evidence suggests that the Australian Government has deliberately harmed the only real prospect the world has of heading off the catastrophes that climate change is expected to visit on the Earth. In preparing this book, I have read volumes of public and private correspondence between members of the greenhouse mafia and between them and the Government. It is truly striking that not once in their commentary have any of them expressed concern about climate change. There is nothing about the fate of poor people in developing countries, no sign of regard for those who may be displaced from island homes, no mention of the potentially devastating effect on the environment in Australia... nothing. Only one issue preoccupies them: how to protect the profits of the fossil fuel-based industries. How do these people square what they are doing with their conscience? Do they ever worry about the appalling consequences if they are wrong?[155]

While I would perhaps consider use of the word "corrupt" to summarise climate policy under John Howard, in describing climate policy under Kevin Rudd, the word that springs to mind is "tragic". In 2007 Kevin Rudd was elected as Labor Prime Minister on a platform of climate change action. He was handed the gift of bipartisan support, which he then squandered through dysfunctional leadership and poor communication. And as Philip Chubb notes in his book *Power Failure*,

---

154 Weihers et al, "Sharing the responsibility of dealing with climate change."
155 Hamilton, *Scorcher*, 226.

The communications void that Rudd created was filled by those who saw their role as being to foster and capitalise on anxiety, fear and doubt. In particular, it came to be filled by Liberal and National Party voices and their allies in the growing movement that doubted science.[156]

During the first few years of his term as Prime Minister, Rudd focussed his energies on chasing industry support, in order to be assured of Liberal support. However in failing to communicate at all with those most in favour of a price on carbon (who were not in favour of the tremendously generous concessions being offered to the coal industry) he disenfranchised some very important players. The Greens and leading environmentalists considered the Rudd government's proposed Carbon Pollution Reduction Scheme (CPRS) as being much too weak.[157]

Somewhat ironically and disappointingly, it was in fact the Greens who blocked the CPRS in the Senate in 2009. In hindsight, they may regret their decision to block this imperfect scheme (a move that was largely based on ideological purity). If the CPRS had come into existence in 2009 without much fanfare or controversy, it might then have been gradually perfected. Despite its imperfections, it would have been more effective than the Coalition's Direct Action Policy!

Political history is full of such pivotal moments. The failure of the Copenhagen Summit at the end of 2009 is another example, and this event had a profound influence upon Rudd, and consequently Australian climate policy:

> Rudd pinned everything on success in Copenhagen, which would save his CPRS and the planet. He believed profoundly that he could influence the outcome, and that the more he was able to exercise leadership, the more likely this was. But the summit to save the globe was always going to fail because of the undisciplined, belligerent and unbridgeable divergence of views among the power blocs represented. When the inevitable occurred

---

[156] Philip Chubb, *Power Failure: The inside story of climate politics under Rudd and Gillard* (Collingwood: Black Inc Agenda, 2014), 66.
[157] Chubb, *Power Failure*, 84.

amid complete chaos, Australian's prime minister was shattered. This was where his descent into personal hell started and where Australia's momentum for climate action stopped and the nation became devoid of a policy – any policy – to price carbon.[158]

By the time it was Julia Gillard's turn to tackle climate change policy in 2010, the national landscape had changed significantly. There was a hung parliament, and the support of the Greens had become more necessary than ever. Unfortunately Gillard made the mistake of calling the emissions trading scheme a "tax" rather than a "price". This became the "great lie" that was seized upon by the Coalition, from which her political career never really recovered. As Chubb articulates well,

> …the new government's task was harder overall. This was because Rudd, having failed miserably to provide any kind of a narrative to explain the need to price carbon, had wasted and then lost voters' commitment to reform. Gillard faced significant voter disenchantment and misunderstanding.[159]

This era saw voters become increasingly ignorant, confused and cynical regarding climate change, with debate regarding the issue fixated on the carbon "tax" itself, rather than the need for action.[160] And this was no mistake, but rather the result of an intentional decision to not mention climate change:

> Signalling the extent of the government's fear, disillusion and disappointment, a mid-March 2012 meeting of ministers decided that to keep talking about climate change was playing into Abbott's hands, so they agreed to stop. In just over four years Australia had moved from a country galvanised by the need to act on climate change to a place where ministers no longer dared even to use the words. While the Gillard Government, against all the odds and showing tremendous determination, had instituted a major climate policy, it now could not say why. In Australian public discourse the term "climate change" had died.[161]

---

158 Chubb, *Power Failure*, 86/87.
159 Chubb, *Power Failure*, 133.
160 Chubb, *Power Failure*, 192.
161 Chubb, *Power Failure*, 233.

In summary, while Gillard was able to achieve success in implementing the Clean Energy Futures Package in 2011 (the four pillars of which were a price on carbon, investment in renewable energy, additional funding for energy efficiency, and improved land use through the Carbon Farming Initiative) it came at such a high political cost to her personally that the cause of effective climate change action was once again pushed back by many years.[162]

During the 2013 Federal election, tactics reminiscent of the Howard era were again employed to cloud the key issues and misinform the public. This includes what I heard when I attended a local 'meet the Candidates forum' organised by the Australian Christian Lobby (ACL) and hosted by a church within my electorate. Various arguments presented by my local Liberal party candidate were along these lines (my paraphrase):

> Australia's emissions only account for 1.4% of the total global emissions, so why should Australians have to punish themselves with this big tax on everything when it won't even make a scrap of difference to global emissions? And why should Australia do anything when China's emissions are growing so rapidly? Australia's emissions will be nothing compared to China's in the next few decades!

These arguments, while superficially appealing to the "polluter pays" principle, disregard the fact that a significant proportion of existing emissions – much of which will remain in the atmosphere for hundreds of years, have already been generated by Australia (rather than China), and our per-capita emissions are the highest among OECD countries.[163] We remain a relatively wealthy nation and accordingly the "ability to pay" principle should be taken into account. Additionally, as Clive Hamilton points out, such an argument is far from pragmatic:

> If the world were made up on only 71 nations, each of which was responsible for 1.4 per cent of global emissions, then no-one would

---

162 Chubb, *Power Failure*, 182.

163 The Australia Institute, "Facts Fight Back: Emissions by OECD countries > Check the Facts", 12 October 2014. http://www.factsfightback.org.au/emissions-by-oecd-countries-check-the-facts, accessed 25 October 2014.

take any action and we would continue to emit greenhouse gases until catastrophe befell us.[164]

More recently Dr Adam Lucas (of the Science and Technology Studies Program at the University of Wollongong) pointed out that our national population in proportion to the global population (approx. 0.3%) now contributes domestic emissions of about 1.8 % of global emissions. In addition, once the impact of burning Australian coal in other nations are taken into account,

> ...plans to triple or even quadruple coal export volumes over the next 10 years would raise Australia's total contribution to global GHG emissions to toward 9 to 11 % by 2020 – an order of magnitude commensurate with that of Middle East oil.[165]

Additionally, the "what about China?" argument that began circulating in the Howard era is becoming an increasingly dicey tactic: China now has emissions trading schemes set up in several provinces and is taking climate change extremely seriously. For one thing, their air is so polluted from the burning of fossil fuels that many of its citizens are finding it hard to breathe! While countries like Australia and Canada are busy trying to rip more fossil fuels out of the land, much of China is busy adapting its manufacturing industries to focus on renewable energy and clean technology.[166] As a concrete demonstration of this trend, "the growth of its electric power system ... is now being powered more by renewables than by fossil fuels."[167]

The *new* argument I heard in the lead up to the 2013 Federal Election was this: "The carbon tax isn't working anyway, emissions are going up". Actually, it's widely known within energy markets that emissions were in steady decline while the price on carbon was in place. More specifically, in the past three years

---

164 Hamilton, *Scorcher*, 31.

165 As quoted by Dr Andrew Glikson, "Methane and the risk of runaway global warming" *Climate Spectator*, 31 July 2013. http://www.businessspectator.com.au/article/2013/7/31/science-environment/methane-and-risk-runaway-global-warming, accessed 31 July 2013.

166 See *When a Billion Chinese Jump: How China will save mankind – or destroy it* by Jonathan Watts for a great (and terrifying!) summary of China's recent environmental policies and developments.

167 John Matthews and Hao Tan, "China roars ahead with renewables" in *The Conversation*, 16 December 2013. http://theconversation.com/china-roars-ahead-with-renewables-21155, accessed 20 February 2014.

there has been a consistent drop in electricity consumption – bucking an upward trend that lasted more than a century (including the Great Depression and two World Wars).[168] Reduced reliance upon electricity from the grid should not come as too much of a surprise, with well over 1 million Australian households now having solar PV installations on their rooftops and a better-late-than-never recognition that the cleanest electricity is actually the electricity you don't use (i.e. a growing focus on energy efficiency and energy conservation).

What my local Liberal Party candidate failed to say when attempting to quote Greg Hunt (the current Federal Minister for the Environment) is that under Labor's carbon price scheme it was *projected* that emissions would still go up between its introduction in July 2012 and the year 2020, which is a completely different matter. Statements regarding carbon emissions need to make clear whether they relate to total or per capita emissions (as population growth may mean that a natural increase in total emissions occurs despite stable or reducing per capita emissions), plus there are some significant emission sources that were not included in the carbon price (including transport and agriculture – which are both significant emission sources). It is important to carefully assess sweeping statements about the impact of the carbon price on total emissions, and the ABC has since drawn attention to the Liberal Party's misleading statements through its FactCheck campaign.[169] The reality is that the Carbon Pricing Mechanism introduced by the Gillard Government did for a range of reasons exceed expectations with regard to its success in achieving emissions reductions (though you're unlikely to read about this in the mainstream media).

Tony Abbott sailed to victory in the 2013 Federal Election, having sworn a blood oath to repeal the carbon price. While he saw his election win as the bestowal of a mandate to this effect, even polling booth exit-survey data suggests that a clear majority of Australians were still seeking strong and effective action on climate

---

168 Hugh Saddler, "The demand drop mystery explained." in *Climate Spectator*, 6 January 2014. http://www.businessspectator.com.au/article/2014/1/6/energy-markets/demand-drop-mystery-explained, accessed 30 March 2014.

169 ABC FactCheck, "What Greg Hunt didn't say about the carbon price and emissions." *ABC News on-line*, 4 October 2013. http://www.abc.net.au/news/2013-10-01/greg-hunt-carbon-emissions-misleading/4989750 accessed 20 October 2014.

change at the time of the election.[170] Many would argue that there were a whole host of reasons for the election outcome, including the rather chronic instability within the leadership ranks of the Labor Party. After a couple of failed attempts, Abbott was successful in passing his repeal legislation in July 2014, with the support of Clive Palmer's new Palmer United Party Senators.

The general direction Tony Abbott has taken action on climate change has been unsurprising: in this regard he has certainly followed through on a number of his election policies. However the speed and breadth of his efforts to dismantle significant achievements of the prior government, including immediately shutting down the Climate Commission and attempting the same with the Climate Change Authority and the Clean Energy Finance Corporation (which had proven itself to be both effective and profitable) have been genuinely shocking to people committed to climate action. Additionally, not only does the Abbott Government have no Department of Climate Change, for the first time since 1931 Australia no longer has a Minister for Science! And in a move that is reminiscent of climate policy development under Howard, early in 2014 Dick Warburton was appointed to head up a review of the Renewable Energy Target. In writing for The Guardian, Graham Redfern had these rather scathing words to say about Warburton and his position on climate science:

> If you look down there at the bottom of the barrel and see the deep gouge marks in the oak, you might find a trace of DNA left by the scraping fingers of Dick Warburton. That's the place where Warburton found his "evidence" that the world's scientists are split on whether or not climate change is being caused by humans. The evidence in question is known as the Oregon Petition - one of the feeblest factoids in the climate science denial hymnbook that's cited almost as often as it has been debunked.[171]

---

170 The Climate Institute, "No mandate for carbon laws repeal, national poll finds", *The Climate Institute: Media Release*, 23 June 2013. http://www.climateinstitute.org.au/articles/media-releases/no-mandate-for-carbon-laws-repeal,-national-poll-finds.html, accessed 26 October 2014.

171 Graham Redfern, "Australia's renewables adviser scrapes the bottom of the climate denialist barrel." *The Guardian*, 25 February 2014. http://www.theguardian.com/environment/planet-oz/2014/feb/24/climate-change-dick-warburton-sceptic-australia-renewable-energy-target-review, accessed 20 August 2014.

To our international shame, political moves like appointing a climate change denier to head up a review of the Renewable Energy Target only confirm suspicions that Australia is returning to its old ways of seeking to wreck any meaningful action on climate change.[172]

Our national political leaders have been toying with this issue for far too long, with little genuine progress as a result. And, to make matters worse, it is now apparent that "history repeatedly shows that there seems to be an inbuilt bias for economic modellers [sic] to overestimate the difficulty and cost of reducing pollution."[173] All this time we have been arguing about the difficulties and costs of action, when the challenges are not nearly as enormous as we've been led to believe.

There is a way forward. It doesn't have to cost us the earth. However, at the time when we really needed astute, courageous, bipartisan leadership on this issue climate change has again been trivialised and used as a political football to confuse, divide and conquer. So much for climate change being "the greatest moral challenge of our time" (to quote Kevin Rudd) – or perhaps this just goes to illustrate how morally ill-equipped we are to deal with such a great challenge!

In an article written for *The Melbourne Anglican,* Former Supreme Court of Victoria Judge David Harper AM reminds us that "the primary duty of governments is to ensure the safety of the governed". While climate change may not involve invading armies, it does present us with a "clear and present danger". Even though it may not be immediate, it does not make the threat less real. He goes on to summarise our national position well:

> The Politicians now in Canberra, of whatever political persuasion, cannot escape history. They must not be allowed to relinquish the obligation of trust by which they are bound. We have all the powers with which citizens of a democracy are endowed. We must ensure that each Member of

---

172 Ian McGregor, "Australian and Canada are leading the wreckers at Warsaw." *The Conversation*, 19 November 2013. http://theconversation.com/australia-and-canada-are-leading-the-wreckers-at-warsaw-20403, accessed 20 August 2014.

173 Tristan Edis, "Reducing pollution – not half as hard as governments think." *Climate Spectator*, 26 September 2014. http://www.businessspectator.com.au/article/2014/9/26/carbon-markets/reducing-pollution-not-half-hard-governments-think? , accessed 20 October 2014.

Parliament allocates primary importance to the reality of climate change and to the adoption of policies which will minimise its impact. It can be done.[174]

It can be done indeed. However with only the Direct Action policy in place we are again losing precious time. And as the window of opportunity for effective mitigation gets smaller, we commit ourselves to harder, more costly, and more urgent work later on.

3.6 A Complacent and Complicit Church

Key Points:

The evangelical church has gotten off to a very slow and rather shameful start when it comes to action on climate change.

While theological progress has been made with many churches making positive statements regarding the necessity of creation care, this has not yet brought about significant change.

We have often marginalised and disenfranchised those within and outside the church who have a deep concern for God's good planet, rather than encouraging and commending them.

Our failure at this level is sin, and the consequences will be reaped by the world's most vulnerable, and by future generations.

---

174 David Harper, "Ensuring our safety is the first duty of government", *The Melbourne Anglican*, May 2014, 19.

So where has the church been through all of this? Unfortunately it has generally been tame and quiet, particularly within evangelical circles where environmental concern has often been associated with *environmentalism, paganism and liberalism* – among other things. While the World Council of Churches, World Vision and the Seventh Day Adventist Church had a presence at the First Earth Summit in Rio in 1992, simultaneously a group of evangelicals gathered to pray that the meeting would be shut down – such was their concern regarding the nature worship and paganism that they associated with the "green" movement![175]

However there have been a few shining lights within our more recent history. In 1970 Francis Schaeffer wrote *Pollution and the Death of Man: The Christian View of Ecology*. Rather than focussing specifically on issues of climate change, Schaeffer wrote broadly of the issues of pollution and the breakdown of our fragile ecological system, expressing deep concern about what was taking place and observing that "the whole problem of ecology is dumped on this generation's neck".[176]

Interestingly, in this book Schaeffer responds in great length to Lynn White's now famous (or infamous) article from 1967 in which the finger of blame for the planet's many environmental woes was pointed clearly at the orthodox Christian doctrine of *domination of nature*. Many environmentalists still hold to Lynn White's views and see Christians, the Bible and the Church as being the underlying cause of our predicament. Francis Schaeffer was one evangelical Christian who responded to White's short article. He called Christians to think, to repent and to respond, and while he conceded that there was some truth in White's argument, it was Schaeffer's firm belief that Christians have a unique call with regard to ecological concern:

> On the basis of the fact that there is going to be total redemption in the future, not only of man but of all Creation, the Christian who believes the Bible should be the man who – with God's help and in the power of

---

175 Loren Wilkinson, 'Christianity and the Environment: Reflections on Rio and Au Sable', *Science and Christian Belief* (5:2, 1993), 143.

176 Francis Schaeffer, *Pollution and the Death of Man: The Christian View of Ecology* (London: Hodder & Stoughton, 1970), 8.

the Holy Spirit – is treating nature now in the direction of the way nature will be then. It will not now be perfect, but it must be substantial or we have missed our calling. God's calling to the Christian now – and to the Christian community – in the area of nature ... is that we should exhibit a *substantial healing* here and now, between man and nature and nature and itself, as far as Christians can bring it to pass.[177] (emphasis mine)

In 1990 a group of scientists from around the globe (including 32 Nobel Laureates and NASA scientist James Hansen) wrote an open letter to the leaders of religious communities, expressing their deep concern for our precious environment and also recognising that the world's religions had a vital role to play:

Problems of such magnitude, and solutions demanding so broad a perspective, must be recognized from the outset as having a religious as well as a scientific dimension. Mindful of our common responsibility, we scientists, many of us long engaged in combating the environmental crisis, urgently appeal to the world religious community to commit, in word and deed, and as boldly as is required, to preserve the environment of the Earth...

> As with issues of peace, human rights and social justice, religious institutions can be a strong force here, too, in encouraging national and international initiatives in both the private and public sectors, and in the diverse worlds of commerce, education, culture and mass communications...
>
> The environmental crisis requires radical changes not only in public policy, but also in individual behavior. The historical record makes clear that religious teaching, example and leadership are able to influence personal conduct and commitment powerfully.[178]

In Australia, it has generally been the Uniting Church and some sectors of the Roman Catholic Church that have heeded such calls and led the way in terms of encouraging Christians to embrace creation care and better understand the important relationship we have with God's Earth. Interest in environmental

---

177 Francis Schaeffer, 50.

178 "Preserving & Cherishing the Earth: An Appeal for Joint Commitment in Science & Religion." http://earthrenewal.org/Open_letter_to_the_religious_.htm, accessed 10 October 2014.

matters has undoubtedly flourished in more liberal theological environments where an emphasis on eco-theology has been encouraged and well resourced.

Environmental concern also flows naturally out of a concern for social justice, which the Uniting Church pursues as a matter of significant importance. In its first public statement in 1977 the newly founded Uniting Church in Australia made its commitment clear:

> We are concerned with the basic human rights of future generations and will urge the wise use of energy, the protection of the environment and the replenishment of the earth's resources for their use and enjoyment.[179]

Australian biblical scholar Dr Norman Habel has had significant involvement in international initiatives such as the Earth Bible Project[180] (which encourages a reading of the Biblical text from the perspective of 'Earth', following a number of eco-justice principles) and the Seasons of Creation, which is designed to be celebrated by faith communities each September. [181]

In December 2006 the Climate Institute published a ground-breaking paper by the title *Common Belief: Australia's faith communities on climate change*.[182] In this document, representatives from a wide range of faith traditions (including Buddhists, Muslims, Jews and Sikhs) express support for action on climate change. Here are a handful of statements from groups within the Christian faith tradition:

> ***Rt. Rev. GEORGE BROWNING, representing Anglicans***
> Holy Scripture reminds us that the earth is the Lord's and everything in it (Psalm 24:1). All of creation belongs to God, not to human beings. We are part of the created order, and our first calling by God is to be stewards of the earth and the rest of creation (Genesis 1:28-29). So when we exploit God's creation to breaking point, we break the most fundamental

---

179 The Climate Institute, *Common Belief: Australia's faith communities on climate change*, (Sydney, December 2006), 8. http://www.climateinstitute.org.au/articles/publications/common-belief.html, accessed 9 June 2014.
180 "Earth Bible", http://www.flinders.edu.au/ehl/theology/ctsc/projects/earthbible/
181 "Seasons of Creation", http://seasonofcreation.com/
182 The Climate Institute, *Common Belief*, 36.

commandment known to us: out of our greed and selfishness, we knowingly cause the degradation of the world's ecosystems instead of protecting the design that issues from the Creator's generosity. Wilfully causing environmental degradation is a sin.

The Christian faith is certainly about personal salvation. But it is more than that: Christianity is first and foremost a concern for the whole of the created order — biodiversity and business; politics and pollution; rivers, religion and rainforests. The coming of Jesus brought everything of God into the sphere of time and space, and everything of time and space into the sphere of God. All things meet together in Him: Jesus is the point of reconciliation.

Therefore, if Christians believe in Jesus they must recognise that concern for climate change is not an optional extra but a core matter of faith.[183]

The Australian Christian Lobby

Within Australia, good stewardship demands that we mitigate the effects of drought and rising sea-levels on our water resources, our agricultural industry, our low-lying coastal cities and our biodiversity.

One of the ACL's main concerns is that the consequences of climate change will be felt most heavily by those least able to bear it. Developing countries, which already struggle with the burdens of poverty, corruption, and natural disasters, are likely to bear the brunt of climate change. We will be looking to the Australian Government to contribute to global action in order to protect this class of people from the consequences of climate change.[184]

The Baptist Union of Australia

The creation teaches us about God (Rom 1:20; cf Job 39:1-42:6). While all things belong to God, God has entrusted the care of creation to humans (Ps 24:1; Gen 1:28-29; 2:15).

---

183 The Climate Institute, *Common Belief*, 8.
184 The Climate Institute, *Common Belief*, 10.

The relationship between humans and the rest of creation is therefore one of interdependence and stewardship. We are creatures shaped by the same processes and embedded in the same systems as those that sustain all other life. Yet as God's stewards we bear an ethical responsibility for the care of the Earth and the welfare of all living things.

We bless God for his greatness and goodness, his mercy and grace, and his love and justice evident in the creation. We enjoy the beauty and pleasures of God's creation. We are sustained and satisfied by its provisions. We are amazed by what science reveals of its structure and systems. We are awed by the miracle of life that continues to unfold day by day.

We also acknowledge that humans have often denied our interdependence with the creation and abrogated our stewardship of the creation. One major result of this is the global environmental degradation and climate change we now face. Overwhelming scientific evidence shows that humans have caused much of the global warming occurring today.

Climate change is one of the most significant threats to our economic and social life. It is imperative that governments and corporations, as well as individuals and local communities, respond to the current global environmental crisis. Failure by national governments to respond to climate change in decisive ways may result in unmanageable cost blow-outs and irreversible devastation to ecosystems and biodiversity. Further, failure to address climate change may ultimately contribute to the suffering and death of millions of the world's poorest and most vulnerable people, and to the forced migration of millions more to cooler and less physically threatening regions such as Australia.[185]

The Catholic Church

Rapid climate change as the result of human activity is now recognized by the global scientific community as a reality.

As pastors of more than a quarter of the Australian population, we urge Catholics as an essential part of their faith commitment to respond to the reality of climate change — with sound judgements and resolute action.

---

185 The Climate Institute, *Common Belief*, 15.

For our part, the Catholic Bishops of Australia offer the hand of cooperation to all spiritual and secular leaders in Australia. We do so in an act of solidarity, knowing that the Earth is our common home. Religion knows the natural world has value in itself. It belongs to God and is only on loan to humans, who are called to care for it. Therefore, the world and all in it must be freed from what can be termed 'a state of suffering'.

Humans are part of the created world, and inextricably part of a material existence. We are indebted to the scientists, environmental activists, rural people, foresters, fisher people, writers, artists, photographers, educators, business people, government officials, society leaders and all who have helped humanity become aware of the dangers of climate change, and created human choices for an alternative future. Such people show that humanity elevates itself when it reaches for a heightened consciousness of Life on Earth.

Future generations should not be robbed or left with extra burdens. Those who are to come have a claim to a just administration of the world's resources by this generation. We need to keep in mind the Precautionary Principle: where there are threats of serious or irreversible damage, lack of full scientific certainty should not be used as a reason for postponing remedial measures.[186]

The Australian Evangelical Alliance, for evangelical Christians

Christians worship God who is Father, Son and Holy Spirit, the creator and redeemer of the universe. God entered into a protective relationship, not only with the people of the world, but also with its other creatures. God gave a particular privilege and responsibility to humanity to tend and care for the world as a participation in divine purposes. One day the whole of creation will be renewed. (Psalm 24; Genesis. 9: 1-17; 1: 26- 31; Ephesians 1:10; Romans 8:21; Revelation 21).

Christ has given the church the task of caring for people and the creation. In regard to large-scale environmental issues, God's call to love our neighbours means taking a global focus. It means recognising that there

---

186 The Climate Institute, *Common Belief*, 18.

is unequal access to natural resources; that the effects of environmental disasters fall unevenly on the people of world. It means understanding the greater difficulty of poorer nations and the moral responsibility of wealthier ones. It means genuinely loving our global neighbours through just, loving and sacrificial action (Matt. 22: 34-40).

There is now no reputable science which denies either that climate change is happening or that a large part of global warming is human-induced. But there is still time to avoid the top range of risk — provided that we do what is necessary and act immediately. For the Australian Government, this would mean establishing a clear policy framework for significantly reducing emissions by the end of the next parliamentary term.

Global warming will affect everyone but its impacts will not be distributed evenly. The wealthy bear more responsibility for producing greenhouse gases, but the poor suffer more from their effects. We must acknowledge our special moral responsibility as a developed nation that has benefited from the causes of global warming by leading the way in finding solutions.[187]

Lutheran Church in Australia

The church recognises that the possibility of unchecked climate change causing irreparable damage to our planet and placing the survival of human life on earth at risk makes the issue of climate change one of the most important moral issues facing humanity.

From a Christian point of view, the destruction of the earth's life-sustaining systems represents an affront to God as the creator of all things and also represents a dereliction of our God-given duty as human beings to care for our world and to use its limited resources responsibly.

Global warming and its consequences involves a failure on the part of humanity to recognise that our responsibility as human beings is to care for each other and to act for one another's good, not just our own. From a Christian perspective it involves a failure to follow Christ's command to "love one another as I have loved you". Our duty to love one another

---

[187] The Climate Institute, *Common Belief*, 21.

applies not only to our own generation but also to all generations to come, whom we may be condemning to death or a life of suffering by our actions.[188]

General Eva Burrows, for the Worldwide Salvation Army

The Salvation Army believes that, as people made in the image of God (Genesis 1:27), we have a responsibility to use the resources of the earth in a way that ensures that people in this and future generations do not suffer from poverty or injustice. This is part of our stewardship of the earth and our love of others. In the modern world, Christian stewardship implies large-scale and permanent changes in attitudes and behaviour towards God's creation, so that we begin to "replenish the earth" (Gen. 1:28).[189]

The Uniting Church

Our commitment to the environment arises out of the Christian belief that God, as the Creator of the universe, calls us into a special relationship with the creation — a relationship of mutuality and interdependence which seeks the reconciliation of all creation with God. We believe that God's will for the earth is renewal and reconciliation, not destruction by human beings. This was expressed as the very heart of our mission in our foundation document:

God in Christ has given to all people in the Church the Holy Spirit as a pledge and foretaste of that coming reconciliation and renewal which is the end in view for the whole creation. The Church's call is to serve that end."

... We regard climate change as a serious threat to the future of humanity and the planet. Our creation of greenhouse gas emissions and our failure to plan for a sustainable future is seriously exacerbating the problems we face. The threat posed by climate change therefore challenges the way we live in a fundamental way. If we are to meet and overcome the challenge we must think creatively about the organisation of our social and economic institutions, our relationship with each other across national and cultural boundaries and our relationship with the environment.

---

188 The Climate Institute, *Common Belief*, 28.
189 The Climate Institute, *Common Belief*, 32.

Some humans consume the earth's resources whilst others pay the price. As one of the major producers of greenhouse gas emissions per capita, Australia must acknowledge that it has a responsibility to address the social, economic and environmental policies which support our continued reliance on fossil fuels. As long as we abuse the atmosphere and entire ecosystems for the sake of short-term economic gain for a few, we undermine our own future. It is important that Australia's social, economic and environmental policies begin to reflect that social justice and ecological justice are not competing interests, but have shared solutions. It makes good economic and political sense to spend money to ensure the long-term well-being of our natural world — there can be no security for humanity without a healthy ecosystem.[190]

While these statements all paint a reasonably clear picture of a Church that is openly committed to caring for creation and acting on climate change, this has unfortunately not been reflected in reality. With the exception of one or two denominations and a relatively small number of specific faith communities, the very serious issue of climate change has continued to be marginalised within the Australian Church. And it has often been the case that individuals or families with strong environmental concern have felt marginalised too, and undoubtedly for the most passionate this would perhaps be reason enough to leave the formal, institutional Church altogether.

It is increasingly evident that climate change is not an isolated issue. In one sense it is a disturbing and distressing symptom pointing to a complex web of problems, with our idolatrous greed at the centre. With an Australian Church that is generally in decline, it's also not surprising that a strong call for repentance and radical change has not been high up on the agenda of the average local Minister, who may simply be trying to tend to the needs of aging, shrinking congregations. When it comes to the climate crisis we now face, the church has largely been both complacent and complicit.

---

190 The Climate Institute, *Common Belief*, 36-37.

It is our conviction that there are several important facets to this tragedy. We begin with these six:

1. Firstly, we have been **unfaithful to our God**: as stewards and caretakers of his good creation, and as those who have been blessed to be a blessing.

2. Secondly, we have **failed in our proclamation of the Gospel** to present a message of reconciliation that includes the mending of our broken relationship with creation.

3. Thirdly, we have **damaged our witness** to those earnestly searching for a hope-filled faith that cares deeply for the natural world. An other-worldly Church that is disinterested in urgent matters of *this present world* can present a significant obstacle to those with deep environmental and social concern.[191]

4. Fourthly, we have **accommodated almost completely to the secular materialism of our era**, including the idolatry of rampant consumerism. We have failed to adequately take to heart Jesus' warning to watch out for greed, and his timeless wisdom that tells us that "life does not consist in an abundance of possessions" (Luke 12:15).

5. Fifthly, we have **rarely presented a holistic and hopeful Gospel** that fully captures people's affections and imagination. Are we genuinely excited about the restoration of all things? Do we really believe that we will witness – indeed see, hear, taste, touch and smell – the renewed Earth, where we finally know the abiding presence of Jesus with us and partake in the wondrous reconciliation of all things?

6. Finally, through wilful ignorance and inaction we have been **neglectful in our love of our neighbour**, particularly of those who will bear the brunt of a disrupted climate: the world's most poor and future generations.

---

[191] In this regard the growing popularity of Eastern religions among Westerners should not surprise us, as the practice of Eastern religions, such as Buddhism, is often far more holistic than a number of popular/contemporary expressions of the Christian faith.

Activists and scientists are seeing that people of faith and religious institutions are a vital part of the way forward. Despite some obvious and significant differences in worldview, many are now seeking to partner with Christians and churches to make a difference. This summary of an article from the Science Policy Forum makes the case very well:

> Humanity is at a crossroads. Do we continue trends of preceding decades that lift people out of poverty and extend life spans, but in the process run down the planet's natural capital? Solutions to this profound problem will require greater cooperation among people. The rise of market fundamentalism and the drive for growth in profits and gross domestic product (GDP) have encouraged behavior that is at odds with pursuit of the common good. Finding ways to develop a sustainable relationship with nature requires not only engagement of scientists and political leaders, but also moral leadership that religious institutions are in a position to offer.[192]

The authors of this article, who work at Cambridge's St. Johns College and the University of California, "believe that next big step is to mobilise the people with the help of the Vatican and other religious organisations".[193] Pope Francis has been vocal in recognising the issue of climate change, resulting in many now looking to him to help lead the way.

---

192 Partha Dasgupta & Veerabhadran Ramanathan, "Pursuit of the Common Good" in *Science*, 19/09/2014, Vol 345 No 6203, 1457. http://www.sciencemag.org/content/345/6203/1457.summary?sid=c72e5e00-16dc-4a83-a0d8-e7f67b01cc6f, accessed 5 October 2014.

193 Reissa Su, "Pope Francis, Catholic Church Key to Climate Change Effort." *International Business Times*, 22 September 2014. http://au.ibtimes.com/articles/567006/20140922/pope-francis-vatican-climate-change.htm#.VDcPt9McRTs, accessed 5 October 2014.

3.7 A Called, Compelled and Courageous Church

Key Points:

There are very hopeful signs that we are observing the dawn of a new era for the Australian Church as she engages with this crucial issue.

Climate change is a pressing issue in its own right, but is symptomatic of deeper issues within our society, including our idolatries and addictions.

Now, perhaps more than ever before, the world needs really good news, and sacrificial love in action.

Many are beginning to reflect on the legacy they will leave behind, as it is distinctly possible that future generations will encounter a lower standard of living than their parents.

The call is there, our God goes before us, and the infrastructure is in place: will we find the courage, the compassion and the love to act now?

It is our deep hope and prayer that the tide is finally turning: Australian Christians are indeed starting to stand up, speak out and take action on climate change, as are other Christians around the world. There are so many signs of hope.

One recent example highlights this. As a part of the 2013 National Day of Climate Action, concerned Christian leaders Mick Pope, Byron Smith and

Jarrod McKenna were all able to speak at the large capital-city rallies held in Melbourne, Sydney and Perth. To have Christians sharing in the public sphere about how their faith informs and compels their commitment to pursue a safe climate represents some significant bridge-building. For a range of reasons environmentalists and evangelicals have historically tended to be generally quite suspicious of each other, so the trust and grace extended in this exchange was really quite remarkable!

Australian activist and Anabaptist Jarrod McKenna provides us with an inspiring and challenging example of what it can look like when Christians take climate change action seriously. In his article *What would MLK do? Christians and climate change* (published on the ABC's Religion and Ethics website) McKenna draws inspiration from the example of Martin Luther King and is uncompromising in advocating for commitment to a holistic Gospel – one that is good news both for the poor *and for all of creation*:

> What our response to the ecological crisis reveals is how ill equipped our Western imagination is to understand the Gospel as good news for all of creation. Much like in the '60s, the issue was not that the Gospel wasn't good news for the oppressed seeking racial justice and reconciliation – it's that too many Christians were blind to see it. They were blind to the reality that the Gospel meant not only personal transformation but social transformation, just as many of us are blind to the Gospel being more than just personal and social, but ecological and cosmological as well.
>
> Polite talk of 'creation care' or 'stewardship' as a kind of appendix to the Christian faith is completely inadequate. The Gospel is not that Jesus came to redeem individuals. The Gospel is that Jesus came to redeem all of creation by grace, and we as individuals, together, can be a witness to that. Our ecological crisis calls us to rediscover the earth-affirming beauty of the kingdom of God… Climate change, like the church's complicity with racism, calls us to repent and rediscover the fullness of the Gospel. Guided by the Holy Spirit, we must return to Scripture and reconnect what should never be separated: the doctrine of the incarnation, the cross,

the resurrection and the kingdom are always connected to the doctrine of God's good creation. To care about the poor is to care about climate change. To care about climate change is to care about the poor. And the Gospel of Jesus Christ, when seen in all its redemptive splendour and beauty, is good news to both![194]

Jarrod McKenna doesn't just write and speak about peace, he lives it out in creative – and at times surprising – ways. His engagement in non-violent action has won him acclaim and respect from a wide range of people around the world, including many activists outside of the church.

Bill McKibben is another inspirational Christian who has been taking the issue of climate change incredibly seriously for decades. His 1989 book *The End of Nature* is considered to be one of the first books written specifically about global warming for a general audience. Though it was a best-seller, and translated into 24 languages, as Bill McKibben reflected in a recent interview with Jarrod McKenna:

> ... my theory of change when I was 27 was that I would write a book, people would read it, and they would change. The first two parts worked, the third didn't really, and in retrospect no shock there: this is the biggest, most difficult problem the planet has ever faced by a large margin. It's the biggest thing human's have ever done, and dealing with it effectively means going to the very heart of the modern economy. And so it is no wonder that we have made scant progress.[195]

With a background in journalism, having worked for organisations such as the *New Yorker* magazine, McKibben tends not to write specifically for Christian audiences, and as such his books don't really emphasise his own faith journey. However during this interview with Jarrod McKenna, McKibben took more opportunity to talk about how his work intersects with people of faith.

---
194 Jarrod McKenna, "What would MLK do? Christians and climate change" *ABC's Religion and Ethics*, 18 January 2011. http://www.abc.net.au/religion/articles/2011/01/17/3114379.htm, accessed 20 February 2014.
195 Taken from a transcript of an interview of Bill McKibben by Jarrod McKenna (recorded in June 2013, for an Iconocast podcast), provided directly by Jarrod McKenna.

Most of the people we work with around the world are poor/black/brown/ Asian/young, because that's what most of the world actually is. And what do you know, they are actually just as concerned about the future as anyone else, maybe even more so since if you're in those categories the future is bearing down harder on you! ... One of the things that has been quite wonderful to see is the widespread participation of a huge variety of faith communities in this world... I can remember 20 years ago when I was first starting this work and one of the things we were trying to do, one of the many things, was stimulate a kind of religious environmental movement, at least in the Christian tradition which is what I come out of. Liberal churches all thought "well the environment is something – a luxury – that you get to once you've dealt with poverty and war and things", and the conservative churches all thought "environmentalism is a weigh-station on the road to paganism" or something: it was very difficult to get anyone involved.

But that has changed, even that first day in 2009, the big day of action, there were so many wonderful things. I remember being extraordinarily happy to see a picture from Wheaton College in Illinois – in the most evangelical liberal arts college in the States, Billy Graham's home – and they were holding a big demonstration. That wouldn't have happened even two of three years ago there. The week before, one thing that made our organising much easier across much of Europe was that the quite wonderful and plain spoken patriarch of the Orthodox Church Bartholomew, leader of 400 million Eastern Christians, gave a sermon in which he said, and I quote, 'Global warming is a sin, and 350 is an act of redemption' – that made it a lot easier! [350 is a worldwide climate change movement to reduce CO_2 in the atmosphere].We had enormous success in, across say India and the Asian sub-continent, across the Muslim world and in places where all these traditions intermingled. I had been to Bethlehem to do some organising a month before, a hard place even to get to really, and the Dead Sea is shrinking dramatically as the temperature rises and so people wanted to do something but there were too many military checkpoints in the way to have a big cooperative demonstration. And so the Jordanians said we'll do a giant human 3 on our shore of the

Dead Sea, the Palestinians said we'll do a 5 on Palestine and the Israelis made the 0 on their beach. It was quite a beautiful demonstration, it attracted an awful lot of attention because for once people were putting aside their other quarrels to deal with problems that really unite us.[196]

In partnership with organisations such as MarketForces, this global movement 350.org has in recent years been pushing hard to see a divestment campaign take root whereby people and institutions withdraw financial and social support of the fossil fuel industry. The influence of the oil industry is particularly strong in the United States and has consistently undermined efforts to address climate change at the level of national policy and regulation. McKibben reflects:

Exxon made more money last year than any company in the history of money: we're never going to have enough money ... to take them on. We're going to need a different currency, and that currency is our bodies and creativity and spirit.[197]

Both McKenna and McKibben have demonstrated a willingness to risk jail time for their commitment to this cause, which is an telling sign of just how few workable options environmentalists feel they have left. McKibben reflects on this tension during the interview:

If we are going to dress and keep the earth as we are instructed to do in Genesis 2 – that means unfortunately going off to Washington and going off to jail and having people run for office and doing what we can to bring down these big mining companies and big oil companies and whatever else. It's a fight and I'm afraid there is no way around it.

It is certainly hopeful and encouraging to see Christians passionately take hold of this issue around the world, and to see this issue unite rather than divide. Interestingly, McKibben has actively encouraged older people to take some political risk in terms of engaging in non-violent action, acknowledging that younger people have more to lose in terms of educational and career prospects.

---

196 Excerpt from recorded interview of Bill McKibben by Jarrod McKenna.
197 Excerpt from a recorded interview of Bill McKibben by Jarrod McKenna.

Many older people are indeed getting involved, fuelled by significant concern about the world their own grand-children will inhabit and what sort of legacy it is that their generation is going to leave.[198]

During the 20th century it was increasingly thought that every generation would continue to experience a standard of living that was better than that of their parents. In fact, I was intrigued to see this quote attributed to the then Health Minister Tony Abbott in December 2004: "Almost never before in the history of the world has there been a happier country than contemporary Australia." [199] Ironically as he spoke those words Jared Diamond was probably writing these:

> [A] society's steep decline may begin only a decade or two after the society reaches its peak numbers, wealth and power. ... The reason is simple: maximum population, wealth, resource consumption, and waste production mean maximum environmental impact, approaching the limit where impact outstrips resources. On reflection, *it's no surprise that declines of societies tend to follow swiftly on their peaks.*[200] (emphasis mine)

We no longer have good reason to presume that a pattern of ever increasing wealth and happiness can be expected for future generations.

> Indeed, we are the first generation of people who now know that our children's grandchildren will indeed not walk the same Earth. They will live on a planet so less hospitable and predictable than it is now that it is unimaginable to us.[201]

Accordingly, in a very real sense, no one should feel the burden of climate change more than today's young adults. Children are perhaps too young to appreciate what a world with a disrupted climate is going to mean for them personally – and they are certainly too young to do much about it without the support of

---

198 The "Walk for Our Grandchildren" march is a great example: dozens of grandparents walked 100 miles from Camp David to the White House in July 2013. You can watch the inspiring Chesapeake Climate Action Network video here: https://www.youtube.com/watch?v=Roq9KYcLYLI , accessed 20 October 2014.
199 Hamilton, *Affluenza*, 133.
200 Jared Diamond, *Collapse: How Societies Choose to Fail or Survive* (Camberwell: Penguin, 2005), 509.
201 Katherine M. Preston, "Accepting the Reality of Climate Change" *Sojourners*, December 2013. http://www.utne.com/environment/the-reality-of-climate-change-zm0z13ndzlin.aspx, accessed 10 December 2013.

their parents! It is today's youth who will be the ones left to lead the world in what look to be very turbulent and troubled times. To borrow a phrase from the musician Sting they are set to inherit a "used-up planet".[202]

As mentioned at the beginning of this section, culture is an essential ingredient in the mix of progress on the issue of climate change. It is unfortunate that Australia increasingly takes its cultural cues from North America rather than the UK. It is unfortunate because the latter nation now has an admirable, tripartisan emissions reduction target of at least 80% by 2050, based on a 1990 baseline.[203] One hopeful sign, however, is that many younger people are now taking the lead and filling the gap in terms of evangelical faith and climate change concern in the United States. GreenFaith Fellow Russ Pierson recently made the following observations:

> Beginning in the Reagan era of the 1980s, with the growth in reach of theologically conservative American televangelists, politically conservative strategists began to focus on hot-button social issues related to individual rights, from the pro-life movement to early steps toward the Defence of Marriage Act. These political stances resonated in churches long associated with "traditional" and "biblical" values who preached a Gospel that had become condensed and homogenized into a focus on a "personal relationship" with Jesus, with a spotlight on individual sins. Slowly but surely, it seemed that being "evangelical" meant being Republican.
>
> As evangelicalism has aged, the visible centre – as often caricatured in the media – has, like the Republican party itself, moved ever farther toward the right, increasingly fused with libertarian ideas. Hence, while Richard Nixon worked to create the Environmental Protection Agency in the 1970s, and George H.W. Bush signed the Clean Air Act into law

---

202   Sting, "All this time" from album All this time, (UK: A & M, 1990). Sting and his wife Trudie Styler founded the Rainforest Foundation in 1989. Back in the 1980s Sting was a kind of "secular prophet" for me. He and his wife Trudie Styler founded the Rainforest Foundation in 1989 "after they saw first-hand the destruction of the Amazon rainforests, and the devastating impact it had on the lives of the indigenous peoples who lived there". Refer to "Rainforest Foundation" website:http://www.rainforestfoundationuk.org/Who_we_are, accessed 13 March 2014.

203   UK Government (Gov.UK), "Policy: Reducing the UK's greenhouse gas emissions by 80% by 2050", last updated 20 October 2014. https://www.gov.uk/government/policies/reducing-the-uk-s-greenhouse-gas-emissions-by-80-by-2050, accessed 25 October 2014.

in 1990, today many of these evangelicals stand with House Republicans who preach job growth offered by Big Oil and Big Coal and demonize any economic or environmental regulation that stands in the way of growth.

Increasingly, however, many US evangelicals are finding a "third way" beyond rote assent to the agenda of either political party. There is renewed attention to systemic sin and an understanding that all of creation is loved and cared for by God—and we as humans are but a part of it. To some degree, there is a "generation gap" in US evangelical churches, where younger evangelicals are leading the way in forging this new orthopraxy.[204]

So, as we move into a critical era in the history of the Earth and of humankind, there is tremendous opportunity for a courageous Church that has a clear sense of calling to lead, to challenge, to encourage and to act. Rather than limiting opportunity for service and mission and straining our already over-stretched budgets, engaging in creation care and climate change action will actually remove significant barriers for the Church as it engages with an increasingly conflicted, desperate and hopeless humanity.

There are plenty of people 'out there' who have a deep concern for all that God has made (including people, cultures, flora, fauna, natural wonders) who at present find the Church inhospitable and, to be honest, incomprehensible. And there are people among those we consider to be the 'least reached' who will connect immediately with concepts of ecological concern and care. Conversely, they would be right to be suspicious of a Gospel of good news that is brought from an idolatrous and indulgent Western Church that hasn't even begun to tidy up its own back yard!

Meanwhile, the Western Church faces its own demise with more congregations closing each week, lacking relevance to and connection with the local communities of which they are a part. The question is: can we reorient ourselves

---

[204] From personal (email) correspondence, 15 December 2013. Rev. Dr. Russ Pierson is an ordained minister, GreenFaith Fellow, and a Certified Sustainable Building Advisor who serves as the Assistant Director of Facilities Management and Planning at one of America's premier green higher education institutions, Lane Community College in Eugene, Oregon. Russ was part of the initial cohort in the Christian Earthkeeping concentration at George Fox University, one of the first such programs offered by an American Evangelical seminary.

and our churches as communities of genuine care and concern for both people and planet?

To conclude, these words from one of Australia's own prophets, John Smith, seem as pertinent today as they were in 1988:

> Some Christians fear that responding to history denies divine sovereignty. God calls us to respond to His grace. I don't pretend to understand where sovereignty ends and free will begins, but I do know this – the Bible does not give us any charter for sitting back in our evangelical armchairs and waiting until Jesus comes back, notching up each new tragedy in society as another of the marks of His second coming.[205]

---

205 John Smith, *Advance Australia Where?* (Homebush West: Anzea, 1988), 223.

Section Four:
Stories of hope
Claire Dawson

Introduction
Fuel your imagination

Those who are involved with climate change action know it can be an extremely difficult. Changing our own priorities, assumptions, expectations and behaviours can certainly be hard, but this is the necessary starting point if we are to have any integrity in calling others to join us on this journey. This personal change is also a crucial part of repentance and obedience.

The next step of engaging with others to see change spread and grow can be equally, if not more challenging. As someone who has a general intolerance for gloss and spin, my natural inclination is to share stories of hard work, closed doors, resistance, and feelings of despair and futility. Seeking action on climate change can be a really hard slog! I'm not sure that this brutally honest approach would be a complete waste of time, as we certainly need to be real about our experiences, our frustrations and the kinds of setbacks that people face. Our aim here however is to inspire others to join in this life-giving journey toward a more holistic existence and faith, and to focus on what is possible – even if it sometimes seems improbable! I want to paint a picture of an attractive and heart-warming Church, and a people who are praying, seeking and working in the hope that God's Kingdom would indeed come on Earth as it is in heaven. And it's not fiction: this is the beautiful Church that I have glimpsed as I have collected these stories from both near and far.

Weaving these stories together has actually been the biggest highlight of seeing this book come to fruition. I have met and corresponded with an amazing, committed, inspired, passionate group of disciples around Australia and across the globe. These women and men are shining – though imperfect – examples of what faithful discipleship can look like in our broken world. People have been extremely willing to contribute to this project, yet they all have been very humble in their estimations of what they are up to. Only God knows the true fruit and value of their labours, but their stories can certainly offer us encouragement and hope. And there are so many stories that remain untold. If I had followed all the leads and chased up every story, this book would never have gone to print. So what you are reading is just a sample to inspire your imagination – the tip of the iceberg, if you like.

There is more, so much more, that God is up to in this important season. This short selection will inevitably fail to do justice to the tens of thousands of Christians who are doing their bit in response to this huge challenge, but it is my hope that it will give you just a taste, and create a hunger for more!

4.1 Grassroots involvement

Key Points:

Rather than being an irrelevant distraction, engagement in climate change and sustainability issues strikes at the heart of fundamental issues of life.

Grassroots involvement in sustainability initiatives is opening up amazing opportunities for strong and vital connection with local communities.

The teaching of Jesus provides incredibly relevant answers to urgent questions being asked by people who are keen to find new ways of "being".

Some argue that environmental engagement is an "either/or" project: that Christians do so at the expense of more important priorities such as mission and evangelism. I want to start by sharing some stories of local initiatives which demonstrate that the choice can instead be one of "both/and".

Climate change action involves addressing some of the most basic and fundamental questions of life and lifestyle: how we eat, travel, and relate to one another and the planet. It is about our hopes and aspirations for ourselves and those we love – particularly the children in our midst. It closely intersects with ethics, and what we believe about the world we live in. It really is all about faith…

Cornerstone Community – Bendigo

In 2009, three families moved into a neighbourhood in a town in regional Victoria. They rented homes within a short distance of each other, close to a local primary school and to the university. Among them were a teacher, a social worker, an electrician, an engineer and a couple of students. They are a part of a new monastic network called Cornerstone Community which has been present in regional Australia since 1978. Their aims are to be and share the Good News with ordinary Australians and disciple those who are ready. In particular they seek to respond to the vacuum being created in rural areas as young people leave for life in the cities. They are not a "church plant": their aim is not to leave a congregation behind. Instead they are a little bit like some of the monasteries of old – a small community of believers whose presence made a concrete difference in a town. Their mission community is self-supporting, with most working part-time. They pool what they earn, and affirm the place of work in God's idea of life. As followers of Jesus they are committed to simplicity. They meet personal needs and the needs of the mission around them from out of their shared resources.

In just five years things have grown amazingly for this little community, and there is now a movement of a couple of hundred people connected in a range of different ways, including hospitality, youth groups, play groups, craft groups, dance groups, community gardens, music nights, and pizza nights. Yet they are quick to acknowledge that for them success has never been about numbers. So how does climate change and creation care fit into the story? Here's what Rose Vincent shared with me.

> *Rose:* The fact that we work part-time, pool our income, rent small homes means that it didn't take long before we were sharing our neighbour's yard to grow veggies. Out of that grew a thriving Community Garden that started with around 15 households in the neighbourhood, and has now moved to a disused church property on the top of the hill. This garden has become a key part of our Committed Company's presence in town. We have been given many opportunities to speak about community development and gardening at the invitation of the local council and community groups. Every second Saturday afternoon there will be around 40

people of all ages harvesting the produce, sharing the workload, swapping skills, swapping yarns, sharing cake, having cuppas and growing community. Researchers from Latrobe University have wanted to discover why people are so highly engaged with our garden. They commented that people obviously come here for much more than gardening!

Another initiative that has grown out of our garden is the Bendigo Community Food Network. This is a space for local community and school garden managers to connect over a meal, to collaborate and to share resources, skills and friendship. It is hosted by a different garden each quarter, and now involves other groups interested in food security, eliminating food waste and feeding the poor.

Some of the university students that are involved in our garden community in collaboration with the local sustainability group have designed a website[206] for locals to use that includes a map of fruit trees that grow along lanes and creek beds that are ready to harvest each season. They have established guidelines for ethical gleaning such as to only take what you need, unless the fruit is perishable (e.g. apricots) in which case you pick it all but distribute it around the neighbourhood.

Some of the key leaders in the student environmental group have been involved in our garden, and three of them have become followers of Jesus. We had the opportunity to get involved in the National Student of Sustainability conference that was hosted by Latrobe Uni last year. Christian faith is generally absent from radical environmentalist circles and is viewed suspiciously as being quite toxic. However our way of life is making the Good News very relevant to these people who are looking for something that makes a difference and that works. What's happening in Bendigo around

206 "Falling Fruit Bendigo", http://fallingfruitbendigo.weebly.com/, accessed 4 October 2014.

our community gives them something they can belong to before they believe.

One young man penned these reflections in response to his involvement with this community:

'When I was younger I was staunchly anti-religious and harboured quite a lot of anger toward many religions of the world, particularly Christianity. This softened over the years and by the time I arrived in Bendigo I had a burgeoning interest in the possible existence of the spiritual realm. Through interacting with the Cornerstone Community I, for the first time in my life, got to know a group of Christians who were actually living their values in a community. I was astounded to realise that many of these values actually corresponded with values that I had! Over time I began to explore why they lived the way they do, why they devote so much of their lives to helping community and what their faith means to them. Eventually, through much hard work breaking down my old beliefs, I saw truth in the existence of God and the Good News. I began to read the Bible and actively learn and practice a Christian life... I am not the only student that has been affected by living in the same as these people either. I can count at least a dozen young adults who have been helped back into the flock by the Cornerstone Community here in Quarry Hill.'

Through school friendships a couple of us joined a few other mums who were interested in starting a Farmer's Market for the town. While it was a long and arduous process, it is now ranked as one of Victoria's best markets and provides a warm, friendly, bustling space not only for collecting local grown organic produce, but also for building community, for supporting local producers, and for celebrating an alternative to mass-produced-trucked-across-the-country-generic-stuff-to-eat. And through the generosity of a group called Gospel Resource, we have been able to purchase 'The Old Church on the Hill' which is now a real community hub. We partner with other community groups and have become a place in Bendigo which promotes sustainability, community, and creativity.

This last weekend we held our first 'Spring Fair' which was an amazing success with a couple of thousand people enjoying the day. At the core is our Committed Company, but the community that hosted this is now much bigger. It includes mostly un-churched people who are experiencing life more like God meant it to be. The money raised is going toward building a community kitchen and feast space alongside the garden, from which we plan to offer so much more to those around us.

These are just a few of the stories emerging from our involvement in this community. We still work part-time, pool resources and live in our rented homes; and we still pray 'Your kingdom come'. And it's exciting to see where He is taking us!

You will come across a few references to community gardens in this book. While not everyone might see an obvious connection to climate change, growing our own food brings us back to the basics of life. We re-connect with the land as we roll up our sleeves, get our hands dirty and witness the miracle of God's ongoing provision for us all. A small seed planted in the right place and nurtured to fruitfulness gives us sustenance for life. How amazing is that? And of course as our climate changes, being out in a vegetable garden heightens our sensitivity to these climatic variations. The effects of droughts and heat-waves become painfully evident as we look upon our precious, frazzled crops. We begin to better understand the implications for those in other places, who have no other means of providing for their families other than what they can grow themselves.

Additionally, large-scale commercial food production is fossil-fuel intensive. Machinery, pesticides, and fertilisers all contribute to ecological damage, as does the transportation of all of this food across our vast country. Growing and eating locally is a great step in the right direction, and there are many people for whom this has been common sense for generations. And depending on how the future unfolds, food production might just become an essential life-skill for us once again. In line with the well known Boy Scout motto, one could call it "being prepared".

It would appear that a new kind of ministry of "fruitfulness" emerges when Christians engage themselves in local food production and pursue a more holistic approach to community mission. There are opportunities everywhere!

Transition Towns

In terms of local community engagement, the Transition Towns (TT) model has much to commend it. Initiated in the UK, the Transition Network now incorporates thousands of local groups scattered all over the globe. Here two Christian women share their stories of involvement with TT in their own communities in South Australia and Victoria. Their stories reinforce that engaging in sustainability initiatives is not something that one chooses to do *instead of* making disciples. Rather, TT communities present a perfect context in which to go deep in one's existing community, journeying alongside others while pursuing practical ways to live better lives. As the stories below demonstrate, it also helps to bridge the gap between the theory and the practice, as people grapple with the reasonably significant changes that are needed to mitigate against – and adapt to – a warmer world.

> *Sally Shaw (Adelaide Hills, South Australia):* Many of us now recognise that the way we live our lives impacts our world and all who live in it, human and non-human, and we are aware that our consumer lifestyles are bringing stress and fragmented relationships. But moving from knowledge to action can be a big jump. Transition Initiatives is a global grassroots movement that has found a way to deal with this jump. They are actively and cooperatively creating happier, fairer and more resilient communities.
>
> This movement believes we have to act together now, because infinite growth on a finite planet is impossible. If we plan and act early enough however, and use our creativity and cooperation to

unleash the genius within our local communities, we can build a future far more fulfilling and enriching, more connected to and more gentle on the Earth, than the lives we have today.

While some Christians may argue that they shouldn't get involved because the movement isn't based on Christian values, this thinking is flawed, particularly as these initiatives do in fact hold many Christian values. Its main value is of caring for God's creation and being responsible for the resources he has given us. Members of the Transition movement may not understand God in the same way Christians do, but many embrace various types of spirituality that involve a deep commitment to care for the Earth. This commitment means they choose to live simply, not influenced by consumerism. They share their resources, care for their neighbour (which includes encouraging local businesses), respect their elders and creatively explore activities such as community gardens and building sustainable houses.

I became involved in Transition Towns because I was deeply concerned about the environment; however I struggled to find Christians or churches who shared this concern. Our TT group is local and has a positive relationship with the local council. As a result of hosting a variety of public film evenings and guest speakers talking on topics such as sustainable housing/living, and growing your own fruit and vegetables, we as a family have been challenged to find ways in which to live more simply. This has included making our home more sustainable, using local transport (we only have one car), growing our own vegetables and caring for some council land.

A TT friend lent us the DVD *The Clean Bin Project* which we watched as a family. My 17 year-old son had taken little interest in recycling prior to this, but had a total transformation as a result. He now can't stand seeing anyone buying or using bottled water, among other things.

While our TT group shares a common bond and vision, the capacity for developing a deep commitment to each other is limited because of people's very busy lifestyles, and perhaps also because we don't share the same foundational values. I do however have a huge admiration for those in our group and for many who attend our public gatherings. They are remarkable in the way they have all 'chosen' to make big sacrifices to enable their lifestyles to be sustainable and demonstrate respect of the earth. And I long for the day when these characteristics are considered 'normal' among God's people.

Jan Down (Maroondah, Victoria): First, a confession. My original motivation for becoming involved in the Transition movement was self-preservation. Back in 2008 I was becoming alarmed at what I was frequently reading in the newspaper about climate change, and what I heard elsewhere about diminishing oil supplies. But when I say 'self-preservation', it was more of a 'we' than a 'me': we humans are in trouble, we need to do something.

It was quite some time later that I began to read and think about these things more consciously in terms of creation care – the fact that this is God's world, that we are meant to be responsible caretakers. And that it is not just humans who are in trouble, but the whole of creation, with eco-systems under threat and so many species facing extinction. I don't remember where I first came across that wider interpretation of John 3:16: 'God loved the *world* so much that he gave his only Son...' The idea being conveyed is that Jesus died to save, heal and make whole the entire creation, not just human beings. This was quite a revelation for me. I now have a growing sense of the preciousness of all created things.

Yet what can the individual do? Transition for me filled the gap between conflicted government and impotent individual. If whole communities can do something together, it might just have an impact.

In general terms, Transition Initiatives raise awareness in their communities about climate change and resource depletion; they are visibly involved in practical activities that help reduce emissions while also building resilience, such as clothing swaps, bike repair workshops or community gardens; and they have some influence with their local councils. Here in Maroondah, in Melbourne's outer-east, we are finding increasingly that we have an advocacy role with our council. This is probably due to the particular mix of skills and interests within our group. We have been involved in Council consultations on future plans for two townships, have advocated for better bicycle infrastructure and presented to a Panels Victoria hearing on a proposed shopping centre.

At the micro-local, we have a food swap which means people can swap not only home-grown produce, jams, eggs and so on, but gardening knowledge, and ideas on sustainable living. This is a very small step towards a localised economy, but reconnecting neighbours has to be a good place to start in building the resilience we need to face the future. We also have people collecting used coffee grounds from cafes and delivering them to schools and residents for use in gardens. Connections with schools and other groups are growing.

We have a long way to go, but being part of a Transition Initiative sustains me and gives me hope. We work hard but we have fun too, and give each other a lot of support. My hope is that churches in Australia will come to understand a 'business as usual' mentality is no longer useful, and that the time has come for living out a different view of reality. Christians, after all, ought to be good at that. We recognize a different King, and we live in the hope of a renewed creation.

Meanwhile in New Zealand…

This inspiring story from across the Tasman Sea provides another great example of creation care working in partnership with community engagement and mission. When the opportunity arose, UNOH worker Dave Tims got alongside his neighbour Raymond Diaz and together they were able to see some remarkable results!

Randwick Park is a suburb in the east of Manurewa (south of Auckland, NZ). The local population of approximately 6,000 people have a strong sense of identity. The outer-suburban community is quite distinct as it is bounded by a stream, a major motorway and farmland. Many of the residents are familiar with the various faces of poverty, and there is a particularly high proportion of children/youth in the area. It's here that David and Denise Tims launched UNOH's first New Zealand Team in 2010.

Historically, the Papakura stream has played an important role within this neighbourhood. One of their elders can recall the area as being lush, with a winding creek and thriving bird population. It even served as a food source for local people who would collect eels, watercress and mushrooms. Then in the 1980s and 1990s the area was developed, and as a part of this process the natural creek was "straightened out". With increasing development, the water became more and more polluted, and people were no longer drawn to the area. It became somewhat of a wasteland, characterised by weeds and rubbish. Instead of being a place of natural beauty, it became increasingly barren and toxic. One day a young boy was playing in the creek and he cut his foot on some broken glass. For local resident Raymond Diaz, the boy's uncle, this was the final straw: something had to be done.

With the assistance of his neighbour, UNOH worker Dave Tims, Raymond started a campaign to get the area cleaned up. They requested permission from Auckland Council to work on the area, and they also applied for grants. As a result of a reasonably lengthy process they ended up receiving $10,000 of equipment and more than $30,000 worth of plants! Not only this, they also managed to gather an amazing array of helpers to assist in two tree plantings. People came representing the local high school, primary school, Indian community group, a youth group, local residents, as well as some members of a notorious local gang. The Department of Conservation sent advisors and tractors and co-ordinated weed-clearing. Older people worked alongside younger people, passing on skills and teaching them about caring for the land. Even Correctional Services sent people to work alongside them in this project, as a part of their community service court orders!

The second larger planting involved over 200 people working over two days, with a big BBQ held each evening to celebrate the day's achievements. Local elders, as well as spiritual leaders from the Manurewa Marae (an important cultural 'hub' serving Maoris living in the South Auckland area) were also involved. Participants were particularly pleased with all that they had achieved, and felt a growing sense of pride regarding their neighbourhood. Others have noticed a new optimism in the area.

Admittedly this is a project that will span generations of Randwick Park residents. It will be 25 to 50 years before many of these trees are mature, and in the meantime there are positive signs that the work that has been started will continue on in a range of ways. For example, there are now plans to improve the area with a new walkway. Older members of the community know that it will be their grandchildren and even great grand children who will benefit most from their efforts. As the trees grow, the native birds will return, and the area may again be lush and beautiful. This is an investment in the future, and these saplings are now a tangible symbol of hope for all of those involved.

…and in the United States of America

Geoff and Sherry Maddock live in a downtown neighbourhood in Lexington, Kentucky. Their home is situated in a context profoundly different from the places where they grew up. Sherry was raised in suburban Atlanta, Georgia, USA. Geoff hails from Yackandandah in northeast Victoria, Australia. They met while studying for M.A.s in missiology at Asbury Theological Seminary in Kentucky in 1998 and have remained there working as missionaries.

Their neighbourhood is fairly typical for the North American inner-city context. Empty, trash-strewn blocks dot the streets, while boarded up homes and discarded shopping trolleys add to a sense of abandonment. Yet, it is also a place with a rich history of African-American agrarianism, arts, and culture and a place where neighbours watch out for each other. Additionally, it is also a place where generations of families have discovered life beyond slavery and oppression after emancipation in the late 1860s.

As Geoff and Sherry made their home in this place to pursue justice and love as they had been taught at seminary, they became increasingly concerned for the emerging ecological crisis. They read everything they could about the changing planet, peak oil, and the industrial agricultural food systems that fill grocery store shelves. They asked themselves, "What can we do as Christian missionaries to live faithfully in the midst of a global crisis like climate change?"

They confess that they were mostly depressed and many times paralysed by the science they were reading. The problem seemed too big to fix and too advanced, even if something could be done. In time, they resolved to practice a careful stewardship with the land and resources they had already been given. The Maddock's home garden simultaneously became a place of retreat, protest, and solidarity. They loved the relative calm of the garden in a busy and often noisy neighbourhood. Growing food represented a protest against corrupt and toxic food systems. And, they found solidarity with neighbours as they shared food and chatted over fences about what was growing and how to prepare raw ingredients for the dinner table. In a place where Geoff and Sherry were outsiders, growing food helped remove barriers formed by differences in race, socio-economics, and educational opportunities.

When the home next door to them burned down and needed to be demolished, the Maddocks, with the help of family and supporters, bought the lot and began planning for a larger garden. This 10th of an acre urban farm is home to chickens, bees, dozens of fruit trees, berry and bramble bushes, as well as flower, herb, and vegetable beds. They work according to the principles of permaculture and don't use any chemicals to control pests or weeds. More than 80% of the food grown is given away to neighbours and countless locals and passers-by visit and enjoy a small patch of Eden.

The Maddocks are the first ones to admit they haven't halted climate change, nor have they mobilized a political movement. They have, however, offered a place that speaks to the beauty and necessity of God's creation. It is a place of creativity and re-skilling, a direct challenge to the broader culture of passive consumption that is largely to blame for the environmental crisis we face.

4.2 Alongside the poor

Key Points:

The voices of the world's most vulnerable are clear in calling for action now. Their natural landscapes and ability to survive and thrive are already being impacted in significant ways.

With escalating damage from extreme weather events, climate change is now front and centre for most aid and development agencies: the link between the two – and the impact on the world's most vulnerable – is now undeniable.

Christian development organisations are responding to this call, and taking serious steps to reduce their organisational carbon footprint.

Interesting new development methods incorporate traditional 'development' wisdom with approaches that assist communities in mitigating against climate change and also adapting to likely consequences.

In 2005 TearFund (UK) published a report with the title *Dried up, drowned out Voices from the developing world on a changing climate*. I can't even remember how I came across it, but I do know that it changed my life. The voices of the world's poor were unequivocal: *climate change is already happening, and we are already suffering*.

Stories from places such as Malawi, Ethiopia, Rwanda, Bangladesh, India, Uzbekistan, Honduras and Peru showed that people were struggling with very similar problems: less distinct seasons, more droughts, more floods, an increase in the frequency and severity of extreme weather events and higher maximum temperatures. The results? Desertification, crop failure, famine, forced migration and an increase in water-borne diseases and malaria, just to name a few. The voices of the world's poorest were crying out, pleading for the developed world to take action. This is seen in the following summary in the report:

> Rich countries were urged to take seriously the need to tackle rising greenhouse gas emissions and were called upon to address the cause of climate change at source by supporting initiatives aimed at reducing such emissions. Specifically they were asked to:
>
> - comply with the Kyoto Protocol (or to sign up to it in the case of countries which have not yet done so, particularly those which have the resources and capability to address the issue of emissions)
> - develop a global solution to tackle greenhouse gas concentrations
> - develop and use alternative energy sources that are environmentally friendly, and make this technology available to developing countries at low cost
> - support education across the world to highlight that our behaviour can have far-reaching effects.[207]

Eight years on, Australia is yet to properly heed this call – in fact we now lag well behind many other countries on this issue.[208] And it seems that the church reflects the inaction of our wider society. I can't help but think of this verse in Proverbs: "Whoever shuts their ears to the cry of the poor will also cry out and not be answered" (21:13). The cry of the poor has been for us to hear, to learn, to act and to solve.

Extreme weather events can wipe out decades of development in hours – or

---

[207] Rachel Roach, *Dried up, drowned out: Voices from the developing world on a changing climate.* (TearFund UK, 2005), 36.

[208] John Conroy, "Australia trailing on 'climate performance' measures", *Climate Spectator*, 20 October 2014. http://www.businessspectator.com.au/news/2014/10/20/policy-politics/australia-trailing-climate-performance-measures, accessed 25 October 2014. This article refers to the Global Green Economy Index, which recently rated Australia as last on global warming leadership. Meanwhile the *Climate Change Performance Index* lists Australia as being 57th out of 58 nations, with only Canada performing worse!

even minutes.[209] It is not surprising that within the Australian Christian not-for-profit sector it is aid and development organisations such as CBM, TEAR and World Vision that are leading the way in responding to climate change. Their workers and overseas partners know first-hand what climate change means to the world's most poor and vulnerable: they are seeing the consequences of climate disruption with their own eyes.

Climate change is undermining their efforts to work with those who most need assistance. What good are seeds and the skills to grow food if the rains never come? What good is a school building that is washed away in flash flooding? Extreme weather events can devastate existing infrastructure and agriculture, and expose communities to immediate physical harm as well as the ongoing threat of contaminated water supplies and limited access to food. Again, action need not be an "either/or" decision in terms of choosing the environment over people, or choosing climate change action over the delivery of aid and development programs. For these organisations to have integrity it must be "both/and": working to alleviate poverty *and* working to minimise negative climate impacts.

In July 2013 CBM produced a report entitled "The Green Office Project". Its foreword includes the following statement:

> CBM is committed to improving the quality of life of people with disabilities in the poorest communities of the world. Climate change and environmental degradation impact the most vulnerable poor nations, communities and families. The world's poorest people, including those with disabilities, who make up 20% of this group, are facing reducing access to clean water, adequate nutrition, fertile soils and growing conditions for agriculture and livestock. In line with its vision and mission CBM is determined to be part of global efforts aimed at improving and protecting the environment and seeking to reduce climate change, including through the lowering of carbon emissions.[210]

---

209 Saba Naseem, "The Smithsonian Institution Announces an Official Climate Change Statement", *Smithsonian.com*, 2 October 2014. http://www.smithsonianmag.com/smithsonian-institution/smithsonian-institution-announces-official-climate-change-statement-180952822/?no-ist, accessed 25 October 2014.

210 (This internal report was produced by CBM International in cooperation with The University of Waikato), 3.

Carbon emission reductions can be particularly difficult for organisations that are heavily reliant upon cross-cultural workers and international partnerships. While some air travel remains essential, they are making significant efforts to reduce the need for emission-intensive travel wherever possible. This includes intentionally clustering meetings (reducing the need for people to make multiple trips to multiple destinations) and utilising teleconferencing technology.

Equally, efforts to reduce emissions associated with the use of office buildings and office equipment are also receiving increasing support. Internationally, carbon footprint reporting is becoming standard practice and Christian NGOs such as the three previously mentioned are all making significant progress in this area. Actions to reduce emissions include improved building usage/design, energy conservation practices incorporating behavioural change of staff and volunteers, and higher levels of energy efficiency for appliances and equipment.

For organisations such as CBM, this has meant providing guidance and support to partner organisations overseas, so that they can also improve levels of energy conservation and energy efficiency within their operations. This makes complete sense, particularly given the findings of their Environmental Stewardship Advisory Working Group Field Consultations (held in Cambodia and India in 2013): "All elements of the consultations pointed to the very strong links between environmental issues and reduced health and wider quality of life problems for people with disabilities and their families."[211]

In addition to focussing on lower emissions at offices and a reduced reliance upon air travel, relief organisations are often able to implement projects that have a range of social, economic and environmental benefits including a climate change mitigation element. In Niger, World Vision has been promoting a resource management system known as 'Farmer Managed Natural Regeneration' (FMNR) with quite astounding results.

---

211 David Lewis, Sirin Atsilarat and Dinesh Rana, "Environmental Stewardship Advisory Working Group Field Consultations: Key Findings And Recommendations" (*CBM International Report*, draft at 30 July 2013), 3.

Unlike other greening initiatives, FMNR uses existing natural resources. Farmers identify residual stems and stumps of trees which are still alive in their fields, and manage them – through pruning and coppicing – to promote regrowth into trees. It is low-cost and low-input technique which farmers can carry our independently with their own tools. FMNR was pioneered in Niger in the early 1980's. One of the driest and poorest countries in the world, Niger has now regenerated over five million hectares, making it the only African country to be experiencing net afforestation. The extent of this expansion highlights another strength of FMNR – its ability to spread spontaneously through farmer-to-farmer interaction.[212]

The results of this process include improved soil fertility, moisture retention and agricultural productivity, as well as opening up new commercial avenues (such as fodder and protection for animals, the sale of firewood, and even a revenue stream generated by the sale of carbon credits). Importantly, FMNR "assists in the mitigation of climate change through carbon sequestration and in adaptation through the regeneration of nutrient-depleted soils and reduced erosion."[213]

Meanwhile TEAR has been working closely with partner organisation UMN (United Mission to Nepal) to achieve positive environmental outcomes. As detailed in one of their recent newsletters, climate change is being taken very seriously by those working on the ground:

> Nepal is the fourth most vulnerable country to the adverse effects of climate change, and the least developed. Glacial lake outburst floods (GLOFs), erratic rainfall, landslides, pandemic diseases, crop failure and hunger all threaten the country. Yet Nepal emits only 0.025% of the total global greenhouse gas emissions. Most people think of low-lying countries as being particularly at risk, but global warming hits mountainous countries hard as well… Climate change has a high human cost, right now. The people living beside the Seti River had their homes swept away as a

---

212 Dr David Lansley, "Economic Development and World Vision: Why, how and what are we doing?" (World Vision Report, July 2013), 9.
213 Lansley, "Economic Development and World Vision", 10.

result of a GLOF in May this year. Fourteen people died, and 50 are still missing. As I write, farmers are waiting anxiously for sufficient seasonal rainfall to plant rice and maize, crops they need to feed their families. Climate change affects everyone, but strikes the poor hardest of all. As a signatory to the UN's Framework Convention on Climate Change, Nepal is committed to help reduce global warming. Nepal has formulated climate change policies which look quite impressive on paper, but in fact there is little action or commitment. All these plans and policies need to be implemented, primarily by the government, but civil society has a huge role to play as well. Hand in hand with communities, UMN is implementing programmes to counter the impacts of climate change, helping build the capacity of partners and communities to adapt to, mitigate and avoid its impacts. We also contribute at national level, raising awareness about the impacts of climate change and advocating for action from the government. UMN participates in national and international networks, forums and working groups, where we offer our expertise and support to make some noise... Without it, the people of Nepal, especially the poor, cannot move towards fullness of life as UMN envisions it.[214]

Specifically, UMN has contributed from its own reserve funds to several environmental projects in order to offset its direct travel and office energy consumption. Their "C-Off" programme, initiated in Rupandehi and Nawalparasi, involves two components:

1. *Installation of Improved Cooking Stoves*: 374 improved cooking stoves were installed in households, after women in literacy groups discussed the problems they had collecting firewood and working in dirty, smoky kitchens. Twelve "stove masters" were trained to make and install these clay stoves, receiving additional income as a result. These stoves saved about 2,190 kg of timber, reducing emissions by 780 tonnes (4 tonnes per stove) and have resulted in cleaner, safer living conditions.

2. *Tree planting:* Women's groups in these two districts planted 581 fruit trees on World Environment Day. Litchis, mangoes and other fruits will help trap carbon, and provide an income when the trees mature.

While all of these initiatives are timely, appropriate and profoundly encouraging, unfortunately it only takes one extreme weather event to wreak havoc within a community. The resulting damage, suffering, and loss of life can be immense, and the recovery process can take years – or even longer.

Typhoon Haiyan, which devastated parts of the Philippines when it made landfall in November 2013, is one recent example of an extreme weather event. Recorded as being among the strongest of Typhoons (and the deadliest in the Philippines: more than 6,000 people lost their lives), it is no surprise that many were quick to point to climate change as a contributing factor. Of course the 2013 United Nations Climate Change Conference was underway within days of the typhoon, so climate change was in the media's line of sight regardless. While some express cynicism in response to those who use such events to raise the issue of climate change (suggesting that they are somehow exploiting the plight of vulnerable people for political gain), in the case of Haiyan the pleas to act on climate change were coming loud and clear from the people of the Philippines themselves, to the rest of the world.

David Lewis (OAM), who works as the Director of Strategic Programmes with CBM, spent time in the Philippines in the wake of Haiyan. Upon his return he made these observations:

> At the end of 2013, I had the privilege of working with CBM in the Typhoon Haiyan zone of Panay Island in the Philippines. It was an extremely busy time, which was often discouraging; however it was at the same time quite motivating as we worked with some of the most vulnerable people, those with disabilities and others affected by this enormous storm. I found it hard to comprehend the destruction which the ferocious winds and huge storm surge had wreaked.

Many families I met in island communities had lost loved ones who had been washed out to sea. Their homes, their fishing boats and nets had been destroyed or lost. Their other main source of livelihood is coconut palms. On many islands up to 60% of these magnificent, tall trees have been pushed to the ground by the typhoon. The trees remaining, will take several years to come into full production again.

While CBM will work together with these communities and other organisations to help rebuild homes and livelihoods, I am haunted by the people's descriptions of this typhoon and the risk that a storm of this size could come again. So many of the older people said they could never have anticipated a storm being that big.

Research indicates that storms in Asia are both more frequent and larger than 20 years ago, and that as climate change continues, this trend will worsen. It is absolutely critical that Christians take climate change seriously. We need to have regard for the creation and also the poorest people who are suffering the largest impact. We need to urgently build the issue of climate change into the 'stewardship of creation', which God has entrusted us with.[215]

---

215 David Lewis, personal (email) correspondence, 14 Feb 2014.

4.3 The triple-bottom line: Profit, People, Planet (3BL)

Key Points:

There are now a number of very 'smart' products and services that provide social, environmental and financial returns.

Christians have been involved in this emerging market, and church communities are often well-placed to support and nurture entrepreneurs and their enterprises in their early stages of development.

Other Christians have taken a new approach, developing products and services that provide social, economic and environmental benefits via the marketplace, rather than through traditional aid or welfare channels. Historically, accountants and investors have focussed on the one single number at the bottom line on financial reports – either a financial profit or loss. In recent decades significant progress has been made in addressing the need for businesses to also provide accountability with regard to their social and environmental impact. The two stories shared here are vastly different, but both demonstrate the significant potential of environmentally sensible products and services.

Barefoot Power

The Barefoot Power story is one that gives me genuine hope. It makes so much sense on so many levels – empowerment, poverty relief, health benefits, and a huge win in terms of emissions reduction:

Barefoot Power Brings Renewable Lighting to Millions Worldwide

In a small village in East Africa, a young boy will finish his homework tonight. And, a mother can work extra hours in the evening to support her family. The fact that this family has enough light to work and can breathe quality air while doing so – with help from a small solar lamp – is an exception.

In developing countries around the world, many light their homes with kerosene lanterns. This cloudy, expensive lighting is dangerous. The open flame and smoke seriously impact indoor air quality and cause countless fires. In fact, more children die from smoke-inhalation related injuries than from tuberculosis or malaria.

But a small Australian company is changing that. Within just a few years, Barefoot Power has successfully brought its solar-powered lamps to approximately four million people in more than 40 developing countries.

The Challenge: Getting Affordable, Renewable Lighting to People

Co-founders Stewart Craine and Harry Andrews launched Barefoot Power in 2005. As consultants in the renewable energy industry, they had witnessed the negative impact of kerosene lighting in places like Papua New Guinea and Nepal. "Power lines go over hundreds of villages on their way to cities, leaving those underneath without light," said Andrews, Director and Co-Founder.

Craine and Andrews wanted to bring renewable lighting to such areas, but existing products were not affordable for those that needed them most. The challenge: develop safe, affordable, renewable lighting and then get it into needy households.[216]

---

216 Adapted from Barefoot Power quarterly shareholders newsletter, October 2014 (as provided by John Altmann).

Impacts of the Barefoot Power products include the following:

- Approximately 800,000 households – encompassing four million people in more than 40 countries – light their homes with Barefoot Power lamps.
- The company is currently achieving sales growth of between 15 and 20 percent annually.
- Barefoot Power has already trained 2,000 entrepreneurs to generate incoming selling lamps and mobile phone charging services.
- Barefoot Power is a 4-time winner of the World Bank's Lighting Africa awards.
- The company now employs 70 people and will generate nearly $ 8 million in the 2014/2015 financial year (approximately double last year). Additionally, Barefoot Power staff on the ground have empowered over 2,000 entrepreneurs in Uganda, Kenya and India with the skills to run micro-businesses selling lamps and mobile phone charging services.
- Barefoot Power products have reduced greenhouse gas emissions in excess of 100,000 tonnes, and would generate carbon credit revenue if a higher carbon price was available. A new round of capital from Oikocredit, the Grace Foundation, Insitor, ennovent and the d.o.b. foundation will help Barefoot Power bring renewable lighting to more of the developing world, with the goal of reaching 10 million people by 2015!

In much of Africa, mobile phones are an essential business tool. Not only are they used for communication, but 'credits' are often transferred as payment for various goods and services in the same way that those in the developed world utilise online banking. But without an electricity supply, many people are highly dependent upon others for the very routine task of charging their phone. In countries like Tanzania, pastors are supplementing their very limited ministry income by offering to charge people's phone for a reasonable fee. They are known and trusted members of the community, unlike some others who are prepared to engage in exploitation or even theft when the opportunity presents itself! So a Barefoot solar system not only provides safe and economical lighting, but also an opportunity for earning income in an ethical and environmentally responsible manner.

Products like this are so exciting as they have the capacity to transform millions of lives across the developing world, and various micro-credit schemes mean that few barriers need exist – even for the world's most poor.[217]

Green Collect

The Green Collect story is another one that excites and inspires me: again there is just so much sense in approaches such as these! While it is less directly involved in climate change solutions, our manufacturing industries use massive amounts of power, and our continuing reliance upon paper products generally adds to deforestation, so efforts to re-use, recycle, and upcycle are all steps in the right direction. The Green Collect model is achieving financial viability as a business model. The environment wins, as do some of those living on society's margins.

In 2002 husband and wife team Darren Andrews (environmentalist) and Sally Quinn (social worker) conducted a feasibility study with the intention of starting a cork collection and recycling business. Based in the Melbourne CBD, they soon began trading under the banner of Urban Seed; a Christian organisation (co-located with Collins St Baptist Church) that works with some of the city's most disadvantaged. Their model is unique on a few levels, but particularly because they offer employment opportunities to those who face significant barriers to workforce participation.

By 2005 Green Collect was able to incorporate separately as a not-for-profit company limited by guarantee, and through the provision of seed funding the organisation has been able to seize new opportunities and develop additional business units.

---

217 A post-script to this story is required for the sake of balance: For many Africans an electricity supply doesn't necessarily mean that kids can read books at night and enhance their educational attainment, but rather than they can plug in a TV and watch the soccer! Another clear example of how Western consumerism has now been exported to nearly every corner of the globe.

Due to the success of the Green Collect model, in June 2012 Social Traders (in conjunction with Sustainability Victoria) prepared a Green Social Enterprise case study that explored the Green Collect story in depth. Excerpts are provided below:

> While Green Collect's distinctive business model has emerged through continuing expansion and innovation, its vision and mission have not changed. Green Collect's commitment to sustainable environmental and social change is realised through its twin aims of:
> - reducing the amount of 'waste' generated and going to landfill;
> - offering new opportunities to people who have encountered significant barriers to employment.
>
> In 2007, Green Collect expanded the range of collected items to include reusable office waste such as furniture and stationery, with the goal of selling the items to other small to medium companies and not-for-profits.
>
> In 2010, Green Collect started to experiment with up-cycling, using collected items to create higher-value products. Today, Green Collect makes notebooks, journals and diaries from unwanted lever arch folders, vintage books and trimmed company stationery.
>
> Today, Green Collect consists of three separate business units:
>
> (1) Resource recovery (office collection services and green office consulting)
> (2) Retail outlets (Yarraville, Dandenong)
> (3) Up-cycling (African Women's Sewing Studio and Elizabeth Street studio)
>
> All profits are retained to expand the business and increase its social and environmental impact. The organisation also has Deductible Gift Recipient status (DGR), which has been critical in helping it secure philanthropic seed funding for its expansion projects.

In 2011 alone, Green Collect collected 65 tonnes of materials, of which 62 tonnes (95 per cent) were diverted from landfill. Since 2002, Green Collect has collected and recycled materials including:

- More than two million corks (7.6 tonnes)
- 6.3 tonnes of toner cartridges
- 31 tonnes of e-waste
- 7700 mobile phone chargers
- 6 tonnes of stationery items for re-use
- 80 tonnes of household items

Green Collect has also conducted 52 waste audits for various organisations.

The [social] impact has been particularly striking at the Elizabeth Street up-cycling studio, which employs formerly long-term homeless people. Green Collect has been uniquely successful in engaging and retaining staff at this work site (located at Common Ground Supportive Housing).[218]

Like the Barefoot Power story, the Green Collect story is unfinished - and with such significant potential it will be fascinating to see how these stories unfold in coming decades. Undoubtedly we will see a significant increase in similar initiatives, which is tremendously exciting!

---

218 Social Traders with Sustainability Victoria, *Green Social Enterprise Case Study: Green Collect*, June 1012. http://www.socialtraders.com.au/learn/dsp-default.cfm?loadref=112, accessed 28 November 2014.

4.4 The Aussie Church in action

Key Points:

The Uniting Church is perhaps a few decades ahead of the rest of the Australian Church in their commitment to caring for God's creation, and there is much that we can learn from them.

The experiences of such churches confirms that engaging in this crucial and relevant space actually opens doors to deeper engagement in society and provides new avenues for local service and mission.

When it comes to the Aussie church, it is the Uniting Church that has very much led the way in calling its members to respond, as well as sectors of the Roman Catholic church. In 2010 a booklet was published entitled *Greening the Church: Australian churches tell their inspirational stories*. It was an initiative of the Five Leaf Eco-Awards Church Project[219] and the Justice and International Mission Unit of Synod of Victoria and Tasmania (Uniting Church in Australia). The foreword included the following statement:

> Church greening is a young and exciting movement. In the United Kingdom, and particularly in the United States, it is rapidly taking off as hundreds of churches and church leaders become engaged in environmental improvement projects and environmental certification schemes. Here in Australia, the movement is smaller, but already there are some really exciting stories coming from churches around the country.

---

219 "The Five Leaf Eco-Awards are a pilot ecumenical environmental change program for churches and religious bodies. The scheme provides assistance, inspiration and recognition for environmental achievements. The program is supported by the Justice and International Mission Unit of the Uniting Church in Australia Synod of Victoria and Tasmania. Designed specifically for churches, The Five Leaf Eco-Awards Church Project encourages faith communities to care for creation and play an effective role in social change towards sustainability." Jessica Morthorpe, Five Leaf Eco-Awards Church Project and the Justice and International Mission Unit, Synod of Victoria and Tasmania, Uniting Church in Australia, 'Greening the Church: Australia churches tell their inspirational stories', (Melbourne, 2010), 32.

This booklet introduces you to some of those stories; but there are many more. The churches whose stories are included in this booklet come from a range of different places, circumstances, beliefs about 'the environment' and motivations for acting; but they share one underlying motivation – a love of God. This love prompts them to care for God's creation, the poor, their children and each other, and challenges them to be good stewards of the earth.

Broad goals of involved church communities including learning more about how to respond to current ecological challenges; practical ideas and resources for living more sustainably; and how to ensure that the world's resources are shared more equitably. There are, of course, a number of common themes in terms of what churches have achieved. These include:

- establishing community gardens (often with strong community involvement and/or artistic themes, and incorporating either worm farms or communal composting)
- bike racks to encourage sustainable travel
- energy efficient lighting and appliances
- replacing old, inefficient heating/cooling systems
- sermons and teaching that are creation-care centred
- film nights to stimulate learning, discussion and action
- getting involved in local environmental projects
- assessing water and energy use of the households associated with a church
- installation of water tanks, solar panels and/or solar hot water systems
- grey water treatment systems
- setting annual emissions reductions targets and making measurable reductions each year
- an Earth pilgrimage (prayer bushwalk) or the establishment of meditation gardens

- acting as a depot for tricky recyclables (printer cartridges, corks, certain plastics)
- reducing paper usage, and switching to 100% recycled paper products
- participating in rallies (such as the Walk Against Warming) as a gathered faith community
- participation in 'Sustainable September' activities

Some stories are worth telling in greater detail, such as the story of **Holy Trinity Church (Tilba Tilba, New South Wales)**. This church community has shaped itself around values of sustainability and creation care, making the most of their idyllic setting which included two acres of land owned by the Diocese:

> Open Sanctuary @tilba is a 'new expression of church' with a particular emphasis on care for Creation. As a Christ-centred community, the sanctuary also offers a contemplative and inclusive way of prayer and gathering from our base at Holy Trinity Church, Tilba Tilba on the NSW far south coast. We believe that the experience of personal intimacy with God's creation may become the necessary metanoia [change of mind, repentance] to energize the individual and communities to act for justice and healing for others and the earth. We offer a space for listening; sharing; and becoming a living model of a community caring for Creation and for disadvantaged people. We also offer a Sabbath space of rest, contemplation and recreation...
>
> In December 2006, Open Sanctuary held a gathering in connection with the Montreal Climate Change conference as part of our commitment to grass roots action. The Sanctuary community felt it was important to honour the many religious voices calling for the moral and ethical imperatives of climate change as part of the conference in Montreal. We were blessed to have poets, musicians and writers amongst us to offer their particular gifts giving expression to a love and desire to care for God's Earth. Since 2006 Open Sanctuary has hosted various speakers and events supporting conservation and action for climate change both within the church and non-church community. ... Many of our congregation

are active in various grassroots eco and social justice issues, and Open Sanctuary has become a place of nurture for people engaged in this work. Our liturgy and prayer are also Creation–focused, and we enjoy the gift of the potent beauty of the natural landscape as part of our 'church'. A spirit of ecumenism and dialogue has grown together with genuine friendships and a clear sense of God's presence amongst us and in that place from which we say, truly, all are welcome. Above all, our invitation is to come and rest awhile to allow the stillness and silence of the place to restore our being.[220]

The story of "A GRAND STAND for the environment" (**Templestowe Uniting Church, Victoria**) is also a particularly interesting one, and it is re-told here:

> Since April 2006, a group of grandparents and seniors from churches in the Manningham municipality have met regularly to share concerns about climate change and its impact upon future generations. We are now an incorporated body known as A GRAND STAND for the environment. We recognize that human beings are standing at an extraordinary moment in world history. Today's grandparent/senior generation is the first to hand on to our children a planet worse off than we have enjoyed. We urgently need to re-think our values and the way we live. We recognize the deep and rising concern about the loss of biodiversity on the planet and the urgent need to protect and restore natural habitat.
>
> We recognize too the profound detrimental effect that global warming has and will continue to have on vulnerable communities in poor countries. With a view to addressing these issues, the group has sponsored public fora in the Templestowe Uniting Church to raise awareness and give opportunity for the voicing of grief, anger, fear and despair as well as striking a note of hope and empowerment. In each of the forums, use has been made of art, poetry, music and ritual, whilst drawing upon the knowledge of and taking inspiration from high quality speakers. We recognize that it is a spiritual/ cultural problem, not a lack of climate science that holds us back in courageously addressing the challenges of climate change. In our fora, we have expressed the challenge of social

---

220 Jessica Morthorpe, *Greening the Church*, 15.

problems and over-arching consumerism in spiritual/ cultural terms when exploring the roots of human involvement in this catastrophic crisis…

Intrinsic to our GRAND STAND philosophy is the importance and place of Indigenous Spirituality. We believe that the gift of Aboriginal relationship with the land is essential in raising our awareness and sensitivity to the suffering of Earth and its peoples. It too, enhances our understanding of the 'sacred' in exploring human identity and our sense of belonging/disconnection to Earth. Thus an Aboriginal speaker/ didgeridoo player has been invited to every forum, and offered a special quality and insight to our presentation… Poetry, music, art, drama, story-telling and ritual woven through the forums have inspired positive response, created awareness and affirmed the spiritual nature of the crisis we face.[221]

Nightcliff Uniting Church (Darwin, Northern Territory) has a deep commitment to issues of social justice, and they have been creatively exploring a range of ways that that they empower others to engage proactively with the issue of climate change, including:

- recognising that their Op Shop facilitates others in reducing their carbon footprints

- establishing a community garden on an under-utilized area of old car park (connections with the broader 'ecological' community have blossomed through this initiative)

- holding a Copenhagen Vigil during the COP 10 process[222] where people could come and express their grief and hope through both silence and liturgy

---

221 Jessica Morthorpe, *Greening the Church*, 7.
222 COP stands for the "Conference of Parties" – which relates to the United Nations Framework Convention on Climate Change. According to the UNFCCC website, "The COP is the supreme decision-making body of the Convention. All States that are Parties to the Convention are represented at the COP, at which they review the implementation of the Convention and any other legal instruments that the COP adopts and take decisions necessary to promote the effective implementation of the Convention, including institutional and administrative arrangements." http://unfccc.int/bodies/body/6383.php, accessed 20 October 2014.

- hosting a movie night in conjunction with the local 'Climate Action Darwin' group

- hosting a Valentine's day 'love the Earth' afternoon tea in conjunction with Top End Transition (a TT group)

- Holding workshops teaching people the art of growing veggies in the tropics.[223]

Courageously and creatively, the **Ecofaith community in Bellingen (Qld)** has committed to meeting outdoors:

> We meet for worship and contemplation out amongst the rest of life, exposed to the elements, to remind us that God is the God of all of life, not constrained to temples made of human hands... Participating on Sundays hopefully encourages and inspires individuals to continue being part of the reconciliation and renewal of all creation during the week. As a group we collaborate with LandCare to "serve and protect" the part of creation where we gather. By meeting outdoors, and in an existing shelter when it rains, and with many people commuting by bike or foot, the ecological footprint of our gatherings is miniscule compared to building a new facility. Our chairs, blankets, and tradewinds tea and coffee are wheeled to the park in perhaps the dodgiest handcart ever made: a testament to the imperfection of evolution. We started meeting in February 2009, and gatherings are open to all people of good will, who want to deepen the connection between Earth (the ecos) and faith, including the implications of Jesus' claim that God is right here, on Earth, amongst and within us, and that we should do something about it.[224]

Meanwhile, **Caloundra Uniting Church (Queensland)** erected 24 solar panels on the roof of their church building – in the shape of a cross! This cross supplies sufficient power for all its lighting and energy needs, including the other groups who use the property. In addition, they estimated that they will generate approximately $2,000 annually by selling surplus electricity to the grid. Community representatives stated that

---

223 Jessica Morthorpe, *Greening the Church*, 19.
224 Jessica Morthorpe, *Greening the Church*, 13.

As we are empowered in our life and mission by the cross, we will be helping to reduce our greenhouse gases by 65% ... We see the sun as a gift from God, as is all Creation. It is our task to use the power of the sun to help create a cleaner environment as we take a small step towards the vital issues of climate change in our beautiful Sunshine Coast.[225]

O'Connor Uniting Church in Canberra has also installed solar panels in the shape of a cross on their building. Another interesting part of their story is the development of a partnership with COGS (Canberra Organic Growers' Society) whereby their disused tennis courts were turned into a community garden! Community grants helped to fund two huge water tanks which will be shared by COGS and the church.[226]

While most of the above stories are from the Uniting Church, there are a growing number of churches from other Protestant and Evangelical traditions that are also taking a serious interest in our relationship with creation. One is the **Cheers Neighbours Network in Banksia Grove** (Western Australia). Geoff Westlake made the following observations about their unique community:

> *Geoff:* We do care a lot about climate change, but just like every other Joe and Joanne Citizen out there, we feel powerless to do much about it. When it comes to sustainability, we are equally concerned about over-population, economic ballooning, peak fossil-fuels, oceanic titration and over-trawling, Gospel justice and stewardship, as well as environmental limits. While I confess to being a pessimist about humans' ability to change before it is too late, here's what we do as a matter of integrity:
>
> - We don't argue with doubters of climate change, but rather enlist them on the common grounds of conservation. Some may disagree that the climate is warming, but even they do agree that we can't keep consuming fossil fuels at this rate. After this point is agreed, the rest is detail. We share our concern to use this current energy frenzy to invent and roll-out sustainable energy infrastructure. We promote solutions when we see them, across our neighbourhood network.

---

225 Jessica Morthorpe, *Greening the Church*, 11.
226 Jessica Morthorpe, *Greening the Church*, 12.

- We suspect that this earth of time and space may be part of the new heaven and earth, or at least play a part for some time yet beyond its obvious use-by date

- We showed films such as An Inconvenient Truth, and Who killed the Electric Car, Story Of Stuff etc.

- We personally know our Federal, State, and Local representatives, and we ask them to legislate about economic speculation; recycling; alternative energy research, development, promotion and installation; fossil fuel dependency; ocean care; justice for the most vulnerable in all this. We make that call fully aware that it will cost us dollars and a lowering of our "standard of living." We consider that we must have serious legislation with serious sanctions, rather than leave the future to "economic forces" which are essentially cannibalistic. Other things include building our house along solar/energy efficient lines, with solar panels that pretty much run our house; fuel-efficient cars; use of Skype rather than travel, where possible, plus meeting locally; use of washable cutlery and crockery; hand-me-down networks for clothes rather than buying new; growing our own vegies and raising chickens, and planting fruit trees in reach of pedestrians; we buy local, small-carbon footprint produce.[227]

---

227 From personal (email) correspondence, 07 August 2013 and also later posted here as "Cheers Environment Review." http://banksia.weebly.com/1/post/2013/08/cheers-environment-review.html, accessed 30 December 2013.

4.5 Alongside the church

Key Points:

Para-church ministries are a significant resource for the existing Church and there are a growing number of organisations that are well-placed to become key partners in the urgent task of climate change mitigation and adaptation.

The development of the Lausanne Creation Care Call to Action in 2012 demonstrates a significant step forward for the evangelical Church.

Hope for Creation and ARRCC (Australian Religious Response to Climate Change) are two key resources for the Australian Church.

Around the world, it is increasingly common to see environmental task-forces or committees within denominational structures or as a part of ecumenical bodies. They resource and empower their members, and often others, to heed the call to care for creation. Additionally, there are now a growing number of organisations that work alongside the broader Church offering support, encouragement and education with respect to the essential place of creation care, including the urgent issue of climate change. This snapshot of just a handful of international and national organisations is in no way a comprehensive list – there are many, many more. It is, however, another beacon of hope. The Christian Church is investing itself in these vital initiatives for the sake of God's precious creation.

The **Au Sable** story begins in Michigan in 1961, when Dr. Harold Snyder decided to establish a youth camp for boys. They quickly realised that the site they had chosen was ideally suited for educational purposes, particularly biology students. Accordingly, it wasn't long before students from nearby universities began attending summer camps too.

A local resident became involved at the site, and after many years of involvement he bequeathed 80 acres of land to the ministry, and by 1979 the Au Sable Trails Camp for Youth has become the Au Sable Institute for Environmental Studies. Dr Cal DeWitt, known to many as the 'modern day father of Christian environmentalism' was the Executive Director for the next 25 years. As their web-site states:

> Today, Au Sable is a highly-regarded educational organization that provides a wide variety of environmental science courses each year to students from nearly 60 participating Christian colleges, with classes in northern lower Michigan and the Pacific Rim, as well as overseas experiences in India and Costa Rica. *Au Sable is a valued resource in educating and inspiring future leaders in Christian environmental science to make a difference in the world.*[228]

Meanwhile, the first **A Rocha** meeting was held near Liverpool in 1983. A Rocha means 'Rock' in Portuguese, and their first project was to look after a coastal area in Portugal which was vital habitat to some important animals and birds. The success of this project over the first decade had resulted in the establishment of national management, and as a result the UK Trust was then able to respond to requests for help from people in other locations. It was agreed that they should focus on "distinctively Christian projects for nature conservation in particularly needy parts of the world".

Soon five new A Rocha projects were developing in Lebanon, Kenya, France, Canada and the UK, with discussions underway in several other locations where there were Christians who were keen to establish "practical projects to express their sense that God cares deeply for his creation".

---

228 "Au Sable", http://ausable.org/

Soon after the turn of the millennium, the pace of growth picked up further, and before long national organisations had been established in several different countries. With new teams having been adopted into the A Rocha family the total number of projects numbers almost 20, as well as small groups of people (such as those in Australia) who are considering new national A Rocha projects. *"A Rocha is an international Christian organization which, inspired by God's love, engages in scientific research, environmental education and community-based conservation projects."*[229]

Meanwhile, founded in 2001 by Christian Ecology Link, UK organisation **Operation Noah** has developed a compelling vision: "the complete decarbonisation of the British economy by 2030." While this goal might seem far-fetched to us, this organisation has been credited with having a significant influence upon the UK's current, ambitious emissions reductions targets. *"Operation Noah is an ecumenical Christian charity providing leadership, focus and inspiration in response to the growing threat of catastrophic climate change."*[230]

Eden Vigil is a relatively new organisation founded by American missionary Lowell Bliss. Eden Vigil (part of the mission organisation Christar)[231] focuses on the reasonably new field of "Environmental Missions", which is also the title of Lowell's recent book, published by the William Carey Library. The main goal of environmental missions is to see creation care fully integrated into cross-cultural mission. As stated in the book, "This then is environmental missions: attending to all four broken relationships, through evangelism, discipleship, church planting and creation care."[232] Or put another way: "An environmental missionary works to restore to rightness the relationships found in broken ecosystems and among the broken people who live there."[233]

---

[229] "A Rocha", http://www.arocha.org/int-en/index.html
[230] "Operation Noah", http://www.operationnoah.org/
[231] "Eden Vigil", http://www.edenvigil.org/
[232] Lowell Bliss, *Environmental Missions: Planting Churches and Trees* (Pasadena: William Carey Library, 2013). 69.
[233] Bliss, *Environmental Missions*, 129.

And, for those concerned about the slippery slope into theological liberalism, these words should provide some level of confidence that the theology behind environmental missions can be simultaneously green *and* thoroughly evangelical: "The crucifixion and resurrection of Christ is the supreme act of creation care and the sole basis for the hope of environmental missions."[234] Lowell Bliss goes on to further expand upon the four broken relationships and the consequences for eternity:

> The current earth on which we live is more than just an airport transit lounge. All evidence suggests that it is the raw material of glory itself. Every instance where we build gold, silver and costly stones into the foundation of restored relationships with God, with self, with others, and with creation will survive the fires of judgement – the dross burnt off – and will pass through, as immortal and imperishable as the body of Jesus himself, the firstfruits of the Resurrection.[235]

Lowell Bliss has also been involved with the formulation of the **Lausanne Creation Care Call to Action**. Those familiar with the history of world missions will realise the significant role of the Lausanne Covenant and the Lausanne Movement. In writing about it, Rev. Tom Houston[236] has written the following:

> That the Lausanne Covenant was agreed upon by 2,300 people from 150 nations from all branches of the Christian church in the space of ten days is one of the miracles of contemporary church history. Some say that if we attempted it now, it would not be possible. ... Humanly speaking, the Covenant was adopted with such wide agreement because it broadened the world view of evangelicals in such a way as to put together in one document an acceptable statement about matters that had been increasingly in tension both in the experience of individuals and in relationships between groups. How this came about was a significant work of the Holy Spirit in our time.[237]

---

234 Bliss, *Environmental Missions*, 76.
235 Bliss, *Environmental Missions*, 104.
236 (Tom Houston was the Lausanne International Director between 1989 to 1993)
237 "About the Lausanne Movement." http://www.lausanne.org/about-lausanne, accessed 20 October 2014.

One of these matters was the relationship between evangelism and social justice, which had been a growing source of concern and disagreement between various branches of the Christian Church.

As a part of the Lausanne Movement, global conferences have been held in Berlin (1964), Lausanne (1972), Manilla (1989), and most recently in Cape Town (2010). In 2012 a group of 57 people from 26 countries gathered in St. Ann, Jamaica to further develop and refine components of the Cape Town Commitment that related to creation care. As this gathering came in the immediate wake of Hurricane Sandy, the importance of the issue was felt particularly keenly. Along with the Cape Town Commitment,238 the full document outlining the Creation Care Call to Action can be accessed (and signed) on-line,239 but a summary of the key features is below:

The Lausanne Global Consultation on Creation Care and the Gospel: Call to Action - St. Ann, Jamaica, November 2012

Primary conclusions:

1. Creation Care is indeed a "Gospel issue within the Lordship of Christ".

2. We are faced with a crisis that is pressing, urgent, and that must be resolved in our generation.

Specific calls for:

1. A new commitment to a simple lifestyle.
2. New and robust theological work.
3. Leadership from the Church in the Global South.
4. Mobilization of the whole church and engagement of all of society.

---

238 Lausanne Movement, "A Confession of Faith and a Call to Action." http://www.lausanne.org/en/documents/ctcommitment.html#p1-7, accessed 20 October 2014.

239 Lausanne Movement. "Creation Care Call to Action: Institutional Signatories." http://www.lausanne.org/gatherings/related/institution-signatories, accessed 20 October, 2014.

5. Environmental missions among unreached people groups.

6. Radical action to confront climate change.

7. Sustainable principles in food production.

8. An economy that works in harmony with God's creation.

9. Local expressions of creation care.

10. Prophetic advocacy and healing reconciliation.

That a collection of evangelicals from a range of different backgrounds have come together to produce such a concise and clear call to the wider Church is a very significant step forward.

Another very encouraging, emerging initiative from the US is **Blessed Earth** which is "an educational nonprofit that inspires and equips people of faith to become better stewards of the earth".[240] Founders Nancy and Matthew Sleeth have an amazing story of coming to faith via their search for answers in response to the environmental crisis:

> One winter break, the Sleeths went on vacation to an island off the coast of Florida. No cars, no roads, no malls—paradise found! They put the kids to bed early and sat outside on a balcony under a blanket of stars.
>
> That's when Nancy asked her husband two questions that would change their lives forever. "What is the biggest problem facing the world today?" Matthew's answer: "The world is dying." No elms left on Elm Street, no caribou in Caribou, Maine. "If we don't have clean air, clean water, and healthy soil to sustain life on earth, the other problems won't really matter."
>
> Then Nancy asked a second, more difficult question: "If the world is dying, what are we going to do about it?"
>
> Matthew didn't have an immediate answer. But he said he'd get back to her.

---

240 "Blessed Earth", http://www.blessedearth.org/

After returning from Florida, the Sleeths embarked on a faith and environmental journey. They read many of the world's great sacred texts, finding much wisdom but not the answers they were seeking. Then one slow night in the ER, Matthew picked up an orange Gideon's Bible. He read the Gospels and found the Truth he had been seeking. Until then, Dr. Sleeth had considered himself a secular humanist; now he became a believer in Christ.

One by one, the entire family followed. And that changed everything— the books they read, the music they listened to, the people they hung out with, and most of all how they learned to love God and love our neighbors by caring for His creation.[241]

A primary activity of Blessed Earth has been the formation of the **Seminary Stewardship Alliance** which describes itself as "a consortium of schools dedicated to reconnecting Christians with the biblical call to care for God's creation." Their stated goal is for "member seminaries to teach, preach, live, inspire, and hold each other accountable for good stewardship practices. The Christian faith occupies a central role in our culture. Seminaries equip, train, and inspire the future leaders of the Church, thereby having a powerful effect across denominations and throughout the world."[242] This vision is exciting, as there is significant potential to transform the attitudes of everyday Christians by influencing emerging leaders within the church. And, there are several Australian seminaries in the process of joining this alliance. However, rather than rely on the next generation of Christian leaders, it is up to all of us to ensure that creation care, and more specifically climate action, becomes a reality that is reflected in God's Church – and soon! Indeed, if it remains simply words and affirmations on paper, then it is hardly a story of hope at all.

On another end of the spectrum, the movement **350.org** is one that is very active outside the Church. It does not operate in any sense as a faith-based organisation. That said I feel compelled to include reference to it here as its

---

241 You can read more of Matthew and Nancy's fascinating story here: http://www.blessedearth.org/about/, accessed 11 October, 2014.

242 "Seminary Stewardship Alliance" website – see http://seminaryalliance.org/

founder, Bill McKibben, seems to be very comfortable talking publicly about his roots as a Methodist Sunday School teacher. We've already been introduced to Bill McKibben in Section Three, but it's worth noting and celebrating the fact that this organisation that was founded in 2007 was just two years later able to mobilise more than 5,000 'actions' in 181 countries![243]

According to their web-site, at the time CNN called this "The most widespread day of political action in the planet's history".[244] Another seven years later 350.org joined with a large number of other organisations in facilitating another massive mobilisation. The centrepiece of the People's Climate March was a massive gathering in New York on September 21, 2014 where over 300,000 people marched to call for stronger action on climate change.[245] This was strategically timed to coincide with the UN Climate Summit which was held the following week, and that same weekend another 2,646 events were held in 162 countries as demonstrations of support and solidarity.

The growing global movement known as 350.org is now poised to be a significant lever in the push to see stronger action on climate change, in part because one courageous Christian decided that it was time to do something. Thank God!

Meanwhile, here in Australia a range of not-for-profit organisations and church denominations have been prepared to take a public stand on this important issue. As already mentioned in the previous section, the **Australian Evangelical Alliance** was one group willing to make a public statement regarding the imperative for Christians to take action and urge for action on climate change. When the Evangelical Alliance launched its '**Ethos**' centre in 2010 the topic chosen for the launch event was climate change, with articulate and passionate presentations from Tim Costello (World Vision), Amar Breckenridge (The Economist) and my co-author Dr Mick Pope (Meteorologist). Since its launch, Ethos has continued to unashamedly speak out on the issue of climate

---

243 For 350.org actions generally include people gathering for a peaceful demonstration of their commitment to action on climate change. In a number of occasions that has involved people gathering together to form the numbers 3 5 0

244 "350.org", see http://350.org/

245 "People's Climate", see http://peoplesclimate.org/

change, with a range of publications, seminars and events devoted to this important topic.

Micah Challenge (a local expression of Micah Challenge International) started in 2004. It is a 'coalition campaign comprised of major Christian aid agencies, churches and grassroots supporters speaking out against poverty and injustice.' A key focus for Micah Challenge is the 8 Millennium Development Goals[246], the seventh of which is to ensure environmental sustainability. This particular goal includes the aim of reversing the loss of environmental resources – something which climate change will make virtually impossible! As such, Micah Challenge has been supportive of action to ensure a safe climate for the world's most vulnerable and for future generations. This commitment is echoed by the wider **Micah Network** in its Declaration on Creation Stewardship (2009). While the whole Declaration is relevant to this conversation, points 5 to 8 are particularly pertinent:

> 5. We confess that we have sinned. We have not cared for the earth with the self-sacrificing and nurturing love of God. Instead, we have exploited, consumed and abused it for our own advantage. We have too often yielded to the idolatry that is greed. We have embraced false dichotomies of theology and practice, splitting apart the spiritual and material, eternal and temporal, heavenly and earthly. In all these things, we have not acted justly towards each other or towards creation, and we have not honoured God.
>
> 6. We acknowledge that industrialization, increased deforestation, intensified agriculture and grazing, along with the unrestrained burning of fossil fuels, have forced the earth's natural systems out of balance. Rapidly increasing greenhouse gas emissions are causing the average global temperature to rise, with devastating impacts already being experienced, especially by the poorest and most marginalized groups. A projected temperature rise of 2°C within the next few decades will

---

246 From the "Micah Challenge" website: "In September 2000, all 189 member states of the United Nations signed on to the Millennium Development Goals (MDGs) – a set of eight targets which aim to halve world poverty by 2015. If achieved, these goals would see the world well on the way to being one in which all people could enjoy wellbeing." – see http://www.micahchallenge.org.au/

significantly alter life on earth and accelerate loss of biodiversity. It will increase the risk and severity of extreme weather events, such as drought, flood, and hurricanes, leading to displacement and hunger. Sea levels will continue to rise, contaminating fresh water supplies and submerging island and coastal communities. We are likely to see mass migration, leading to resource conflicts. Profound changes to rainfall and snowfall, as well as the rapid melting of glaciers, will lead to more water stress and shortages for many millions of people.

7. We repent of our self-serving theology of creation, and our complicity in unjust local and global economic relationships. We repent of those aspects of our individual and corporate life styles that harm creation, and of our lack of political action. We must radically change our lives in response to God's indignation and sorrow for His creation's agony.

8. Before God we commit ourselves, and call on the whole family of faith, to bear witness to God's redemptive purpose for all creation. We will seek appropriate ways to restore and build just relationships among human beings and with the rest of creation. We will strive to live sustainably, rejecting consumerism and the resulting exploitation. We will teach and model care of creation and integral mission. We will intercede before God for those most affected by environmental degradation and climate change, and will act with justice and mercy among, with and on behalf of them.[247]

It is not surprising therefore, that a key response of Micah Challenge has been its ongoing support of the **Hope for Creation** project, which emerged through the partnership and support of Micah Challenge, Ethos, World Vision and the Uniting Church. From its website:

> In 2011, thousands of Australian Christians joined with Christians around the world to pray about climate change, and to call on world leaders to take strong action to combat global warming... We long to see Australian Christians at the forefront of a movement to bear witness to the abundant

---

247 "Micah Network 4th Triennial Global Consultation on Creation Stewardship and Climate Change: Declaration on Creation Stewardship and Climate Change", July 2009 – see http://www.micahnetwork.org/resources/advocacy/declaration-creation-stewardship-and-climate-change, accessed 20 October 2014.

love of God in creation and to walk with Christ in a life of discipleship that defends the rights of the vulnerable poor in the face of a disrupted climate. *The Hope For Creation* campaign features a set of resources to raise awareness, speak gently to the sceptic in your life, take practical action, and advocate to politicians for effective responses to the challenge of climate change and pray.

Catholic Earthcare Australia - an agency of the Catholic Bishops Commission for Justice and Development – has articulated its mission as follows:

> to help promote understanding among people that Creation is sacred and endangered, and must be protected and sustained for present and future generations yet unborn. To play its part in helping protect the integrity of creation and the health of Earth's inhabitants and life supporting ecosystems, Catholic Earthcare Australia is mandated, through the activities of research, education, advocacy and outreach, to give leadership in responding to Pope John Paul II 's call for an "ecological conversion" of the world's Catholics.[248]

Catholic Earthcare, while being particularly active within Catholic circles, is at present undoubtedly the largest (and best resourced) Christian eco-group in Australia, and even more so after being awarded a Federal grant of more than $1,000,000 to form of a *National Energy Efficiency Not-for-Profit (NEEN) Network*. The NEEN Network Project will inform and support energy efficiency awareness in the Australian not-for-profit sector. According to their website,

> The purpose of the *NEEN Network Project* is to educate, resource and connect small to medium sized, not-for-profit organisations across Australia in their efforts to reduce energy consumption, energy expenditure and ultimately their carbon footprint. The *NEEN Network Project* will provide members with free access to energy efficiency information (tailored to the needs of the sector), innovative audit tools/ resources to support change at an organisational level and the opportunity to join a community of their peers with whom they can share ideas, strategies and successes.[249]

---

248 "Catholic Earthcare Australia" website – see http://www.catholicearthcare.org.au/
249 "Catholic Earthcare Australia" website – see http://www.catholicearthcare.org.au/

More broadly, **ARRCC** (Australian Religious Response to Climate Change) supports the entire faith sector in Australia, which includes a significant number of Christians and churches. ARRCC was incorporated in 2007 and their Vision statement is as follows:

> The Australian Religious Response to Climate Change (ARRCC) envisages our nation embracing a sustainable future, one which is based on a more ethical understanding of human prosperity and the flourishing of all. To help create this, ARRCC aspires to influence religious communities of all kinds and all across Australia to actively reflect religious values in their lifestyle choices.[250]

Organisation Chair, Thea Ormerod, continues the story as follows:

> Our vision was essentially two-fold. On the one hand, we wanted to encourage communities across Australia to embrace lifestyle options which would reflect caring for the earth. These were not to be portrayed as some kind of self-sacrifice, but as alternative options which hold the promise of better health, slowing down the pace of life and increasing a sense of connection with the earth and each other.
>
> Eating less meat, championed by the Hindus and Buddhists, is not only important in reducing methane emissions, but is better for ones health and an expression of respect for other life forms. Walking, riding a bike or using public transport similarly means less stress and more exercise. Pursuing energy efficiency can mean lower power bills and, most importantly, nurturing a sense of connection with Creation is spiritually enriching.
>
> Secondly, ARRCC wanted to add another, broadly-based faith voice to the climate debate in Australia. Over the years we have published a number of open letters, made submissions, written our own letters and visited Parliament. A high point in the advocacy work was a Youth Embassy on the lawns of Parliament House in 2012... [Another] high point was the recent launch of our Christian Climate Action kit, which is an online, up-to-date comprehensive resource for local communities to integrate

250 "ARRCC" website - see http://www.arrcc.org.au/

228 A CLIMATE OF HOPE: CHURCH AND MISSION IN A WARMING WORLD

Creation Care into their common life. In a context where the public's support for climate action is lessened and there is a misinformation campaign being waged, Australia needs prophetic voices. ARRCC makes a modest attempt to be one of these voices, to keep faithful to the urgent need to protect the integrity of creation for future generations.[251]

So there are already plenty of people who have well and truly commenced the journey of leading the Church toward a 'greener' future, both in Australia and around the world. The infrastructure is there and much of the hard theological work has already been done. Additionally, people like Thea have put their words into action, demonstrating even a willingness to be arrested.[252] Now to see God's people rise up with passion, compassion, conviction and a clear sense of call!

---

[251] Personal (email) correspondence, 30/07/2013.

[252] Thea Ormerod and a handful of other religious leaders were arrested in March 2014 at the site of the Maules Creek mine (Leard State Forest, NSW). An article by Thea's appeared in the *Sydney Morning Herald*, 12 March 2014, http://www.smh.com.au/comment/faith-and-science-combine-as-religious-leaders-join-fight-for-maules-creek-20140312-34m6h.html, accessed 11 April 2014.

4.6 Inspiration has a face and a name

Key Points:

There are hundreds – if not thousands – of inspiring stories that could be shared. We need to allow our light to shine. The world needs hope.

We must remember that there are wonderfully committed, concerned, and generous people outside the Church doing a whole lot to make a difference.

Don't underestimate the power of your own life and example. God can work profoundly through the words and actions of people – including you!

There are hundreds, perhaps thousands, of stories that could be shared if I delved into the tales of everyday heroes, doing what they can to make the world a better and more sustainable place. For Christians, there is a fine art to this kind of story-telling as we like to be humble about our own achievements, aware that our small actions are but a drop in the ocean. Additionally, Christians are keen to acknowledge the significant role that God has played in calling, equipping, providing and encouraging along the way. The glory is ultimately His.

But while we're instructed not to let our left hand know what our right hand is doing when it comes to giving to those in need (Matthew 6:3), we're also instructed to let our light shine before others that they might see our good deeds (Matthew 5:16). In fact, Ephesians 2:10 says that "we are God's handiwork, created in Christ Jesus *to do good works*"; in Hebrews 10:24 we are told to "consider how we may spur one another on toward love and good deeds". James 3:13 echoes these sentiments: "Who is wise and understanding among you? Let them show it by their good life, by deeds done in the humility that comes from wisdom." Those who care for God's creation should not be afraid to lead by example, to show by their deeds that their faith is genuine (James 2:17). Indeed, some "pagans" have probably been right in accusing us of environmental wrongdoing, but wouldn't they be pleasantly surprised to see a growing number of Christians living admirably good lives among them, consistently performing good, green deeds of active creation care? (1 Peter 2:12)

We need to remain mindful though, that when it comes to caring for creation, Christians are certainly not the only ones capable of good works. Indeed, many of the most inspirational people and stories will be found well outside the walls of the Church. Some have been doing all that they can for many, many decades – and to our shame the Church has rarely supported them, thanked them or encouraged them.

Liellie McLaughlan is someone I have connected with in collecting stories for this book, and she mentioned how inspired she was by someone she met recently, someone who is not at all "churchie":

> Carol is in her mid-fifties. While she is a vivacious and bright mum, her arms are full of scars from years on drugs. She now lives in a housing trust house, working three jobs. She is clean and seeking to make a better life for her and her son. Carol saved her hard-earned money with the hope of going on a family-holiday, yet once she'd saved her $5,000 dollars she decided instead to give back – to give back to the world and community that had supported her when she was unsafe and unwell. So Carol bought solar panels for her housing trust house (and in doing so had to fill in a 30-page document to assure Housing SA that she would not reclaim the money if she had to move)! Carol said: "I just had to do it: I just had to do the right thing. I could not go on this holiday with my son while fossil fuels are making the world a dangerous place."

> Carol's story inspires me. Her story reminds me of the saying: if you want to live sustainably, you have to put in more than what you take out. God the Creator put in so much more than I ever can – I am determined to keep on challenging myself to follow Carol's example: to act because it is the right thing to do.[253]

I wanted to include Carol's story, as re-told by Liellie, as it presents a genuine challenge to us all, with our various excuses for inaction - including that it is too hard, too expensive, too confusing or too complicated. And sometimes we genuinely underestimate our own capacity to inspire others toward action, just as Carol has inspired Liellie.

Years ago a friend told me that she thought of me every time she put something in the recycling bin. Initially, I reflected that this told me more about her than me: that she obviously lacked connection with many environmentally aware people – for surely recycling is just common sense?! There is probably some truth in this, as recycling is now a very normal thing to do in many, many homes. But it also serves as a reminder that inspiration usually has a face and a name. We need people to model a new way of life for us. We need to be discipled in the art of creation care. I have had a similar experience myself, and I now nearly always think of a particular friend whenever I consume various animal products (which is a practice I try to do less often, and now with much greater awareness that an animal has actually given its life for my own dietary indulgence). While veganism might seem like a real stretch for many people, Bron's story highlights just how transformative her decision has been for herself, and for others in her midst:

> My unwitting journey to veganism began one fateful day when my curiosity finally got the better of me and I decided I wanted to know exactly how animals were slaughtered. This was after reading an article in The Age about a burger from McDonalds. I think something in me always knew I would eventually come to a point that I would educate myself about the effects of my food choices; I now just wish I had done it sooner!

253 Liellie McLaughlan, personal (email) correspondence, 19 November 2013.

It's not really rocket science that slaughter is an unpleasant business, but the production line manner in which these sentient beings were being dealt with just shocked me. However many head of cattle, sheep, pigs etc needing to be "deconstructed" per hour, per worker inevitably leads to many animals being many steps down the production line while still being alive and conscious.

I had opened a can of worms that I could not shut and seen things that I could not un-see. I read and watched clips for hours that afternoon researching more and more. The fact that cows must have a calf to lactate never occurred to me and made me feel very stupid indeed: I was a mother, I knew how the whole pregnancy, birth and milk coming in cycle worked – how could I have not wondered before about what happened to the calves that were born? I learnt about male bobby calves who are not economically viable to the dairy industry. They are taken after birth to slaughter with both mother and baby grieving for each other. I learnt about the sexing of newly hatched chickens, where the males (who again have no economic worth) go one way on the production line to be ground alive or suffocated, while their sisters go another way to be used for egg production or meat. These were living, feeling creatures we were using as no more than production units in an assembly line.

As someone who liked to think of myself an aware, ethical woman and an animal lover, I was shocked beyond belief. Life has never been quite the same since. I felt for the rest of that first day like I was being remoulded from the inside out by my Maker. His beloved creation was being so terribly treated and exploited in a way he had never intended, and for the first time my eyes and spirit were opened to the weight that His creation was crying out and groaning under. A lifelong burden for them was placed on my heart.

I let my shocked family know about my immense awakening. One by one my family had their own personal ethical awakenings: first my eldest son told me he wanted to go totally vegan, then a few months later my youngest son did the same. Then my husband ended up becoming vegan when he started reading a book called *Eating Animals* by Jonathan Safran Foer.

Since we as a family of believers (and now also vegans) have embarked on this road less travelled, I feel spiritually lighter and proud of the kinder choices that we as a family make day to day. Voting with your money is a very powerful feeling once you know the truth about something. I think as a family we have a lot more intentional awareness, integrity and authenticity than we did before. I know a plant-based diet is MUCH kinder to the planet environmentally and leaves a much smaller carbon footprint. I would also prefer that the world's grains go to feed starving people, instead of cattle in feedlots being made ready for slaughter.

One of the truly greatest things that has happened since I first went vegan 5 years ago is the wonderful, compassionate people we have met through our veganism. We have made great lifelong friends. We as a family are also on-shore volunteers with Sea Shepherd Conservation Society helping feed the all-vegan crew, whom we admire and respect so much for the front line conservation work they do for the oceans.

We have also had quite a few friends become vegan through us which has really blown me away and been very encouraging. I think more and more, ethically minded people are wanting to know the real story behind what they eat and are starting to question the impact their food choices are having on issues such as the environment, people, food distribution globally, native and indigenous habitats, the oceans – and of course the animals. These animals, be they farmed, fished or native, are so greatly affected by our food choices. And many would be surprised to know that as a family we eat really well and have expanded our food base instead of lessening it![254]

So, as you contemplate what you've read in this book, and decide to take action in some new ways, don't ever underestimate the importance of your own personal example upon those who cross your path and who witness your life. Rarely does the abstract inspire us: it is our respect for a particular person, or place, or story, or tree, or animal, or bird – or the God who made them all – that will ultimately compel us toward action. Allow yourself to be challenged and inspired, and be prepared to be someone who in turn challenges and inspires others.

254 Bronwyn Adderley, personal correspondence, 21 August 2013.

Conclusion
Next steps together in a season of opportunity

There is so much happening in this season. Against a disturbing economic, social and political backdrop there are people rising up in response to what is clearly one of the most significant challenges humankind will ever face. And committed Christians are increasingly among them. Those who work in the field of climate change, or those who just happen to know the facts and take the rather dire projections to heart, are becoming increasingly desperate in their attempts to find appropriate solutions to this crisis.

The necessary response is going to require massive action at every level of society, by everyone who is able. One tremendous opportunity for the Church is that those involved in non-faith-based spaces are seeking out people and organisations to work alongside in moving forward – including Christians and the Church. It has taken far too long to come to this point, but we are gradually coming to terms with the huge challenge before us. In taking the next steps, wherever possible we should explore opportunities to collaborate with others in order to maximise our effectiveness. This should certainly include partnership with those who act toward the same ends but for different reasons.

In my own experience, when those working for environmental action within the secular sphere come across ecologically concerned Christians, they have demonstrated an increasing willingness to talk, listen, share and to trust. This race to save our fragile planet is going to be a marathon, and those with a passion for a safe climate recognise the race must now be run by those best equipped to succeed. We have no time left for failed attempts and second-best efforts, or for the wasted time and energy that result from turf wars and power games.

The Christian Church has infrastructure, she has purpose and she has good intention. She has leaders of character and influence, and she has people infused with the Spirit of the Living God. Just imagine what a blessing we could be to the nations if we could step up and share the role of leading the world toward a safer, more just and more hope-filled future?

In this section we have heard stories from people who dream of such a day, who are working in partnership with God to see the Kingdom come, here on Earth, as it is in heaven. Take some time now to think about which story has most inspired you to make some changes in your own life. Think about other people in your life who have inspired you in similar ways. List the lifestyle changes you'd like to make soon. Make some time to talk with those who might be impacted by these decisions, and think of people who might be able to support you, encourage you and hold you accountable.

Section Five:
What on Earth do we do now?
Claire Dawson

Introduction
The call to live in "climate truth"

Many Christians who are passionate about addressing the challenge of climate change can tell their own story of eco-conversion. I can recall reading Tim Flannery's 2005 book *The Weather Makers* and being "cut to the heart" as a result. Tears, contrition, godly sorrow and repentance followed. While the book I was reading was not written from a Christian perspective, much of it still rang true and demanded a response. I knew I needed to learn more, to re-configure my worldview and to make some practical changes as a result. However it was such a big transition for me that initially I didn't quite know where to start. And so I began my journey of coming to appreciate the vulnerability of God's creation, and the need to respect and care for it, particularly in relation to climate change.

I have certainly not arrived at a point where I can say I live with complete integrity in this area – far from it in fact! As I wrestle with what it looks like to be a disciple of Jesus in this busy and complex 21st century, I am painfully aware that all sorts of compromises still exist in my life. The more I learn, the more I realise there is still much that needs changing about the way I live, the way I view creation, and the way my lifestyle impacts upon all that God has made.

No doubt some aspects of my response to the issue will be worked out over the rest of my lifetime. I see challenges everywhere, but increasingly I also see opportunities. Putting time, prayer and effort into this book has been a part of my response. And the process of reading, thinking, discussing, learning and writing has in itself been a catalyst for moving me forward in pursuing change.

The particular expression *"cut to the heart"* is taken from Acts 2:37, where the crowds hear Peter present the Gospel just after he receives the Spirit at Pentecost. They are on the receiving end of a fairly confronting home truth when Peter tells them, "Therefore let all Israel be assured of this: God has made this Jesus, whom you crucified, both Lord and Messiah." Wow. What a message! And what on earth do they do in response? They are "cut to the heart" and ask "Brothers, what shall we do?"

They knew this important news demanded a response. Life had to change as a result of this stark message that defined a profound new reality. Now I'm certainly not trying to suggest that facts about climate change are as important, life-changing, or history-shaping as news of the resurrection of Jesus Christ. Yet, the reality of human-induced climate change is a big message, especially when one begins to consider the possible ramifications for Earth and all its many inhabitants. This rapidly unfolding situation threatens the survival of all kinds of precious species; plants, animals and human populations on every continent of the globe are increasingly at risk, and it can now almost be guaranteed that our children's children will grow up in a very different world to the one we have known.

So, it is indeed right to ask "what we can do?" as things certainly need to change. Behavioural change is a crucial component of repentance, along with a change of heart and change of attitude. Interestingly, this is exactly what Peter instructs these listeners to do in response to his Pentecost message: "Repent and be baptized, every one of you, in the name of Jesus Christ so that your sins may be forgiven". He calls people to a new way of life.

When it comes to climate change, there are a whole range of things we can do in response. Some of these things you will know a bit about already, through various campaigns run in recent years, or through modelling by friends or relatives. Reading the stories of hope in the previous section might have stimulated your imagination too. There are now millions of people who are passionate about seeing climate change slowed significantly so that its effects are minimized, and there is an abundance of resources to help us take action.

While we could, at this point, choose to launch ourselves into a frenzy of activity, as Christians it is important to first bring all these things before God. It is good and appropriate for us to express our sorrow, our grief, and our concerns. We need to be honest about how we feel and what we think. Doing this will help us live in "Climate Truth" – something that New York-based Climate Psychologist Margaret Klein thinks is a vital part of our response:

> Our society is living within a massive lie. The lie says, "Everything is fine and we should proceed with business as usual. We are not destroying our climate and, with it, our stability and our civilization. We are not committing passive suicide."
>
> The lie says we are fine—that climate change isn't real, or is uncertain, or is far away, or won't be bad enough to threaten humanity. The lie says that small changes will solve the problem. That recycling, bicycling, or closing the Keystone Pipeline will solve the problem. The lie allows people to put climate change in the back of their minds. To view it as someone else's issue—the domain of scientists or activists. The lie allows us to focus on other things. To proceed with business as usual. To be calm and complacent while our planet burns...
>
> The lie says that there is no crisis. That business as usual is fine. That our species is not marching towards its doom. The lie is our enemy, and our survival depends on fighting it. But knowing the truth isn't enough. To beat the lie, we have to do more than know the truth. We have to *live* the truth. We have to act on what we know to be true.[255]

---

255 Margaret Klein, *Living in Climate Truth* (Section I), 27 August 2013. http://theclimatepsychologist.com/?p=156, accessed 10 October 2014.

5.1 Godly Sorrow

Key Points:

Sorrow is an appropriate response to the truth about climate change.

We are culpable, and we are vulnerable. This feels uncomfortable.

Repentance and prayer are healthy initial steps on the journey.

In his second letter to the Church at Corinth, the Apostle Paul writes about *godly sorrow*. He is aware that he has written to them previously with great frankness, and that his first letter hurt them and caused them sorrow. But in the end it was not something he regretted because it was a fruitful sorrow:

> Even if I caused you sorrow by my letter, I do not regret it. Though I did regret it—I see that my letter hurt you, but only for a little while—yet now I am happy, not because you were made sorry, but because your sorrow led you to repentance. For you became sorrowful as God intended and so were not harmed in any way by us. Godly sorrow brings repentance that leads to salvation and leaves no regret, but worldly sorrow brings death. See what this godly sorrow has produced in you: what earnestness, what eagerness to clear yourselves, what indignation, what alarm, what longing, what concern, what readiness to see justice done. (2 Cor 7:8-11a)

Certainly the context is very different, but the process of truth-telling, followed by godly sorrow, repentance and an earnest, genuine response remains relevant for us today. We live in a materialistic, 'loud' culture where we are bombarded daily by messages instructing us to pursue personal happiness as our primary goal. Other than understandable grief over loss of life or other tragedies, sadness is commonly judged as a sign of personal failure or weakness, or alternatively presumed to be a sign of ill-health, such as depression. For Christians this can even be exacerbated as we are told that the joy of the Lord is our strength, and that joy is a fruit of the Spirit – so we should be happy, right?

Yet in the Old Testament there are repeated references to tearing robes, wearing sackcloth, putting ashes on heads, fasting – and even public demonstrations of wailing and weeping during a season of mourning. And this seems to be exactly what God seeks:

> The Lord, the Lord Almighty,
>   called you on that day
> to weep and to wail,
>   to tear out your hair and put on sackcloth.
> But see, there is joy and revelry,
>   slaughtering of cattle and killing of sheep,
>   eating of meat and drinking of wine!
> "Let us eat and drink," you say,
>   "for tomorrow we die!" (Isaiah 22:12-14)

> "Even now," declares the Lord,
>   "return to me with all your heart,
>   with fasting and weeping and mourning."
> Rend your heart
>   and not your garments.
> Return to the Lord your God,
>   for he is gracious and compassionate,
> slow to anger and abounding in love,
>   and he relents from sending calamity. (Joel 2:12-13)

We live in a different age and many of these things may no longer be meaningful expressions of our grief, pain and regret, however we now have very few legitimate options left – particularly when it comes to sharing our grief in public settings. What does it really mean to "rend our hearts?" Sometimes having a little cry just doesn't do our situation or our emotions justice!

Climate change is a complex issue which is intricately intertwined with our politics, economics and our very way of life. Some of the possible consequences are quite dire, and while the responsibility for emissions and action is certainly shared, there is no way to escape the fact that we have all contributed in our own way.[256] We may feel regret for our own actions (or inaction), and grief for what it means not only for us but for those we love and those we are to call our neighbour. There might be a loss of hope, an erosion of trust in those who had a duty of care to govern and lead more diligently, even fear about what the future holds.

If in coming to better understand the reality of climate change, you feel overwhelmed with sadness, regret or sorrow, it would certainly be helpful to explore ways to deal with these emotions constructively and meaningfully, rather than trying to ignore or repress them. In training a new generation of Climate Leaders, Al Gore recently likened despair to denial, stating that we don't have time for despair. I can understand his sentiment, and the need to guard against a despair that descends into hopelessness. However, facing the reality of our situation and the breadth of the emotions that can surface in response is really important. I genuinely believe that God values this honesty, and to grieve over what grieves his heart is a very healthy thing.

Poetry, artwork or song might open doors for appropriate self-expression. For others, journaling, prayer, or a time of stillness and solitude (ideally somewhere outdoors where you can see, touch and smell the creation we're talking about) might also help you process your thoughts and feelings. It can also be appropriate, I think, to repent for collective sins on behalf of larger groups of which you are

---

[256] (at least if we live in a Western, developed nation!)

a part (e.g. family, church, state and nation). Many are now acknowledging the very real need to share and acknowledge grief in more public ways, whether it be simply through talking with friends, or a formal ceremony that enables the public expression of lament and confession.[257]

This might feel wrong to you, particularly if you have been raised to 'keep your chin up' or – to recall Bobby McFerrin's famous song – "Don't worry be happy"! This is one of the vital contributions that faith can actually make to this issue, so don't skip it! Contrition is not something we tend to hear sermons about very often, but it is something that is profoundly valuable in the sight of God:

> For this is what the high and exalted One says— he who lives forever, whose name is holy: "I live in a high and holy place, but also with the one who is contrite and lowly in spirit, to revive the spirit of the lowly and to revive the heart of the contrite. (Isaiah 57:15)
>
> Has not my hand made all these things, and so they came into being?" declares the Lord. "These are the ones I look on with favor: those who are humble and contrite in spirit, and who tremble at my word. (Isaiah 66:2)
>
> My sacrifice, O God, is a broken spirit; a broken and contrite heart you, God, will not despise. (Psalm 51:17)

From Wikipedia:
> **Contrition** or contriteness (from the Latin contritus 'ground to pieces', i.e. crushed by guilt) is sincere and complete remorse for sins one has committed. The remorseful person is said to be **contrite**. It is a key concept to Christianity.[258]

There is no need to fear emotions of despair, or feelings of remorse. As even Wikipedia tells us, this is a fundamental part of the Christian faith. Indeed the grace of God demonstrated in Jesus Christ can then change us and free us, and the resurrection of Jesus Christ gives us immense hope that a new day awaits. We have been forgiven, and we have hope!

---

[257] Some might even feel that specialised support services are appropriate at some point, and Psychology for a Safe Climate might be a good place to start – refer to http://psychologyforasafeclimate.org/

[258] Wikipedia contributors, "Contrition," Wikipedia, *The Free Encyclopedia*, http://en.wikipedia.org/wiki/Contrition, accessed 14 October 2014.

Embracing Vulnerability (Repentance and Prayer)

If you're reading this you're obviously literate, which already puts you at a distinct advantage when compared to millions of other people in the world. Despite a lot of progress in recent decades, there are still 11 countries in which adult literacy rates remain below 50% (all of which are in Africa). Chances are that you also have a reasonable education, a capacity to engage in various forms of paid or unpaid work, and the opportunity to exercise choice in a range of areas of your life (e.g. what to eat, what to wear, where to go, who to relate to). You are probably used to being able to repair some things, or fix some situations when they go wrong; to find answers to at least some of your questions, and to make some progress in achieving some of your dreams or goals.

This is not the case for a significant proportion of the world's population. Many who live in ongoing material poverty lack the basic choices that we tend to take for granted. They lack power. They are not used to solutions, to answers, or to progress. Millions of people spend their days trying to survive and trying to help their loved ones survive. I'm sure for many there are also aspects of their lives that are rich and full in ways we cannot appreciate or understand. But if we were to explain the science and politics of climate change to them, I doubt their response would be to try and fix things: they are used to feeing small and to feeling like they are at the mercy of much bigger forces and systems; without a voice; without a vote; and perhaps without much hope.

For some of us, embarking on the creation care journey can stir similar feelings. We suddenly feel so small, so powerless, so vulnerable, and so very out of control. These feelings can be foreign and scary, especially for those of us who like to be in the driver's seat! I am a recovering perfectionist. I like things neat and tidy and if possible wrapped up in pretty (recycled) paper with a nice (pre-loved) bow. So I can certainly resonate with how uncomfortable this world can be. Fear of the unknown, anger at the injustice of it all, sorrow for those who are already suffering, and frustration that so little is being done when we have known so much for so long! I know from my own experience that there is a tendency to want to rush through this crucial stretch of the road. But it is here that we can grow compassion, develop some degree of empathy and have our hearts genuinely moved and changed. The humbling experience of sitting (or kneeling!) in our powerless vulnerability might even be something that God uses to help turn our hearts of stone into hearts of flesh.[259]

A huge part of what has put us into this environmental mess is our lust for power, control and domination. To keep things simple, we could call this all "sin". Developing humility and a right sense of our place in God's precious creation can be an important part of our healing and restoration. Feeling vulnerable, dependant and helpless is often uncomfortable, but isn't it in our weakness that God's power is made perfect (2 Cor 12:9)? Plus, I have a sneaking suspicion that if the trees really could speak or the rocks cry out they would express similar feelings of helpless exposure too – how very vulnerable is creation right now? Perhaps this is part of creation's groaning and travail that we read about it in Romans 8? And are we not ourselves a very precious part of God's creation too?

This is the place in which we often find ourselves in prayer – perhaps even subconsciously. When there is no obvious solution; when things are just too much of a mess to fathom – we pray, we cry out to God. Perhaps it will just be moans and groans, along with the rest of creation, as we all await our final redemption. But the place of prayer, and of sitting before God with our big, unanswered questions, is an important one. And prayer is certainly one of the primary 'weapons' that we should take into any battle. Richard Foster, well-known for his teaching about prayer, says it well:

> Now I have no naïve optimism that we are going to solve every problem and eliminate all injustice – that must await the return of Christ. But we have a mandate to invade this present evil age and establish beachheads of light and hope everywhere. And we are promised that the gates of hell will not be able to withstand the onslaughts of the Church. To know that we are dealing primarily with a spiritual reality also gives us important clues

---

[259] E.g. Ezekiel 11:19

as to our strategy. The work of prayer, for example, should not be shoved off into some pious corner to be used only in connection with devotional concerns. Far from it! When we confront structural evil we do so in a power drawn from divine resources.[260]

And the Lausanne Creation Care Call to Action ends appropriately with its emphasis on the need for prayer:

> Each of our calls to action rest on an even more urgent call to prayer, intentional and fervent, soberly aware that this is a spiritual struggle. Many of us must begin our praying with lamentation and repentance for our failure to care for creation, and for our failure to lead in transformation at a personal and corporate level. And then, having tasted of the grace and mercies of God in Christ Jesus and through the Holy Spirit, and with hope in the fullness of our redemption, we pray with confidence that the Triune God can and will heal our land and all who dwell in it, for the glory of his matchless name.[261]

So be still, breathe, and know that God is God. And so begin with prayer.

- Pray for *conviction*: that we would see our sin for what it is (particularly our greed and indifference) and embrace a new way of life
- Pray for *forgiveness*: for ourselves, our church, our nation
- Pray for *conversions*: that people everywhere will be reconciled to God *and* to all that he has made
- Pray for *creation*: that it might be substantially healed in this generation
- Pray for *change*: that our hearts, culture, politics, economics and infrastructure would change faster than our climate
- Pray for *courage*: for the Spirit to embolden people into compassionate action and take a stand against the powers, and that they would find wisdom, strength and perseverance in the Spirit

---

260 Foster, *Celebration of Discipline*, 166.

261 "Creation Care Call to Action." http://www.lausanne.org/en/documents/all/2012-creation-care/1881-call-to-action.html, accessed 7 January 2014.

Climate Change Prayer

Holy God,
earth and air and water are your creation,
and every living thing belongs to you:
have mercy on us as climate change confronts us.

Give us the will and the courage
to simplify the way we live,
to reduce the energy we use,
to share the resources you provide,
and to bear the cost of change.

Forgive our past mistakes
and send us your Spirit,
with wisdom in present controversies
and vision for the future to which you call us
in Jesus Christ our Lord. Amen.

© The Anglican Church of Australia Trust Corporation.
Used with permission.

5.2 New ways

Key Points:

We act because we love God. Caring for his world is the right thing to do.

We act because repentance involves change, and where possible righting some of our wrongs.

We act because inaction presents danger and harm for our neighbours.

We choose to change now, rather than wait to have change forced upon us.

While there are certainly a whole range of actions we can take in response to climate change, more than anything we need to appreciate that it ultimately involves a new way of relating to this planet that God so lovingly made. We read in Acts 22 about "followers of the Way" – an early term used for those who chose to follow Jesus.[262] We desperately, urgently need to find a new way – or many new ways, each depending on our different contexts! This might mean that our lives start to look increasingly different to those around us – even to those we call fellow Christians. It might actually mean that our lives have more and more in common with others who share a similar passion for caring for this planet, our home, despite holding a whole different set of values or religious beliefs to us. While this might feel strange at first, in time you might be pleasantly surprised by a new-found freedom. You might even begin to feel more at home in your skin – and indeed on this Earth – than you ever have before.

From the outset we need to keep in mind that there are three reasons for our actions. The first reason is that *we love God*. We treasure what he has made, and

---

[262] It was actually some time before followers of Jesus would become known as Christians, and this is not a term that Jesus ever used with respect to his disciples (refer Acts 11).

we want our lives to honour him. The created world has value because God created it and values it. We can still see that it is indeed good, in so many ways, despite the damage we have done. Our world has value because it is our home, and because it sustains us, those we love, and those we don't know but who have become our neighbours in this globalized world. We breathe this Earth's air; we drink its water, we eat its food, we take shade under its trees. We delight in the songs of birds, we are amazed at beautiful vistas and are overwhelmed by vast oceans and ancient forests. So first and foremost we act for the earth because it is good and right to do so, and because it honours and pleases God.

Secondly, we also act because this is a part of what *repentance* looks like. I still think Lionel Basney said it well two decades ago:

> The difficulty, in ecological terms, with the idea of personal redemption is not that it focuses our attention on our sins. The problem is that the range of these sins has been too narrow, and forgiveness offered on terms too cheap. We have confused the free offer of grace with an escape from responsibility – setting our guilt behind us with ignoring the harm we have done. But restitution means that part of repenting is attending to the harm that we have done. As far as the earth is concerned, we have a serious repentance to perform, and the restitution will be long, complicated, strenuous, and expensive.[263]

Finally, if one attempts to summarise the teaching of Jesus, he ultimately called people to a love for God that is demonstrated through *love for others*. When Jesus is grilled and tested by the religious leaders of his day, he manages to make a great summary of all the laws and rules that had accumulated over many centuries of Judaism. He said:

> 'Love the Lord your God with all your heart and with all your soul and with all your mind.' This is the first and greatest commandment. And the second is like it: 'Love your neighbor as yourself.' All the Law and the Prophets hang on these two commandments. (Matt 22:37-40)[264]

---

263 Lionel Basney, *An Earth Careful Way of Life*, (Downers Grove: IVP, 1994), 119.

264 Elsewhere in Matthew we read something similar, where Jesus says "So in everything, do to others what you would have them do to you, for this sums up the Law and the Prophets." (7:12)

So, the third reason we act is because we choose to love, and our love compels us to try to make a difference in this fragile world. We want to lighten people's burdens and make difficult lives more bearable. We want others to know something of the peace and security and abundance that we have known in our lives – things which God intended everyone to enjoy. We want to lessen the incidence of floods, famines, and fires. We want to fight against disease, disaster and death. We want to be *pro-life* in every possible way because God loves life, and he calls us to join him in this great love. And so it is that we also act in the hope that we will make a difference, but not always being assured that we will.

There are strategic opportunities to pursue change in ways that are more likely to generate real results. We do need to focus on these things for the sake of the world's poorest, as it is only through strategic, focused, collective effort that we can ever hope to achieve measurable reductions to our emissions and in doing so lessen the consequential environmental damage. It will take significant action to reverse the concerning trend toward a warmer planet which among other things is characterized by wilder, warmer and less predictable weather as well as the threat of rising sea-levels.

The damage done already is so significant, and the challenges ahead are so huge, that small incremental changes are not going to be enough on their own. As Bill McKibben says, it is "time to stop changing light-bulbs and start changing systems."[265] We don't have time for a "slow graceful evolution to a new world": the pace of change will be uncomfortable, and the consensus is that it's going to take political action to make it happen.[266]

Yet we need to make the small incremental changes as well, because it is good, and right and faithful, and because sometimes these small actions help to change us as well as those who witness our lives. Importantly, some of these changes will also be important as we seek to adapt to a more unstable climate. Many are now becoming aware that we can either choose change now (which can feel

---

265 Bill McKibben, *Oil & Honey*, 14.
266 Bill McKibben, *Oil & Honey*, 109.

hard) or have change forced upon us later (this will certainly feel even harder)! We need to invest more of ourselves in local and more sustainable sources of food, water, and energy because when disaster strikes, the resources of one's local community often become vital.

There are, as previously mentioned, endless resources available to help you to reduce your own environmental impact. While we will certainly make sure we include some of the most obvious and important suggestions in the following pages, I am particularly keen to highlight some of the things that we can do uniquely *as followers of Jesus*. I believe that the resources of faith open new doors and make all sorts of things possible.

5.3 Getting started

Key Points:

Start by transforming the way energy is used by those in your household.

Share the journey. Travel together. Reconsider your aviation habits.

Switch to renewable energy: buy Green Power and consider rooftop solar.

Get energy aware: use it less, use it smart.

Begin at home

One of the most important things we can do is to join with others in seeking change. Change is hard, but even more so when you are unaccountable and on your own. We worship a Triune God: Father, Son and Spirit. Together they have modelled life in community for us. Isolation is not meant to be a part of Christian experience. So the most obvious place to start is in our own households. This may refer to a family unit, but ideally it will increasingly include all sorts of creative arrangements of living in community and sharing resources. Indeed, the rising incidence of single-person households is one of a number of trends that undermine efforts to reduce our collective carbon footprint. Sharing is one of the obvious keys to consuming less, and when we live alone we generally don't share as much. People living together more regularly share meals. They share warmth in winter and cool air in summer. Proximity also means that people are more able to share cars, fridges, tools and TVs. Meanwhile, single-occupant homes add to urban-sprawl, increase demand for new housing and new stuff, and inevitably mean that more people live further and further from the amenities of city centres, which in turn generally adds more vehicles to the road and pollution to the air.

I used to despise the trend toward bigger and bigger houses until one day I heard someone say (my paraphrase) 'There is nothing wrong with big houses *if* they are full of people.' So true! The issue is not big houses per se, but big houses with very few occupants. In 2009 Australia sat at the top of the list in terms of average house size (214.6 m^2). This is twice the size of the average Australian home in the mid 1950's and almost three times the current size of homes in the UK![267] Big empty houses are black holes of consumption: they require so much extra heating, cooling, furnishings, cleaning and maintenance, but much of it is unnecessary if the space is largely un-used. And what do we do with the extra space? Usually fill it with stuff we don't even need! Additionally, large homes sometimes come at the expense of green space, gardens and trees. Trees play a vital and often over-looked role in the life of our neighbourhoods[268] and garden space is of course good for growing home produce including herbs, vegetables, fruits and even nuts!

267 Greenlivingpedia contributors, "House size comparisons," *Greenlivingpedia: for a sustainable future*, http://www.greenlivingpedia.org/House_size_comparisons, accessed 26 October 2014.

268 PlanetArk's document "The Benefits of Trees" is one on-line resources which explain the numerous social, environmental, health and financial benefits of trees: http://treeday.planetark.org/documents/doc-752-ntd12-the-benefits-of-trees.pdf, accessed 25 October 2014.

If you do happen to live on your own, consider whether this is a necessary arrangement for you longer-term. If you really feel that it is, then get together with a few other neighbouring households and share the sustainability journey with them. The journey is so much more possible, enjoyable and meaningful when it is shared with others, and when the going gets tough it's certainly no good to be alone. To again quote Bill McKibben:

> When people ask me where they should move to be safe from climate change, I always tell them anyplace with strong community. Neighbours were optional the past fifty years, but they'll be essential in the decades to come.[269]

Church communities are a natural place to make such connections. Ask around, you might be pleasantly surprised by who else is interested in pursuing a more sustainable way of life. However if those in your church community aren't yet on the same page, you will undoubtedly find like-minded people through a local community garden, or some other local environmental initiative.

Sharing: old is new

Sharing isn't a new idea. In fact it's a good old one that needs to be re-discovered. In the book of Hebrews we read: "And do not forget to do good and to share with others, for with such sacrifices God is pleased." (13:16) And if that's not enough, consider this instruction from Paul to Timothy:

> Command those who are rich in this present world not to be arrogant nor to put their hope in wealth, which is so uncertain, but to put their hope in God, who richly provides us with everything for our enjoyment. Command them to do good, to be rich in good deeds, and to be generous and willing to share. In this way they will lay up treasure for themselves as a firm foundation for the coming age, so that they may take hold of the life that is truly life. (1 Tim 6:17-19 NIV)

In the Church we're reasonably familiar with the concept of giving. You give, and it is gone. It has been given. Sharing, however, can be somewhat trickier.

[269] Bill McKibben, *Oil and Honey*, 40.

Sharing can take more patience, more involvement, more risk and more grace. I wonder if that is why so many of us are challenged and a bit scared of the verse in Acts where we read about the lifestyle of some of the earliest Christians: "All the believers were one in heart and mind. No one claimed that any of their possessions was their own, but they shared everything they had." (4:32)

Have you been to your local library recently? Where I live, not only do we have a fantastic local library full of great books and DVDs, we also have one of the nation's largest toy libraries nearby. Instead of buying books, DVDs and toys for our kids (that they are often 'over' within a week or two) we try to just borrow and then return. These are structured, facilitated examples of sharing. Our local council has even started semi-regular 'clothes swap meets' where you can exchange clothes you no longer wear (or fit!) for something new and different. Imagine what creative opportunities for sharing our church communities could come up with if we put our minds to it?[270]

Getting around

One of the other areas in which we can make a big difference to our environmental footprint is by sharing transport. While the goal should be to wean ourselves off our fossil-fuel-guzzling cars, if you lack many suitable alternatives there are certainly things you can do in the meantime. This might include two or more drivers sharing a car, giving and receiving lifts, or utilising a car-pooling initiative. The choice of a smaller, more energy efficient vehicle is also a small step in the right direction. With rapid improvements to battery technology, electric vehicles are soon going to become a more feasible option for many.

Ideally, for many of us, sustainable travel should also mean using the bus or train, cycling ... and walking! Again, this is where the rubber really hits the road (pardon the pun). These modes of transport can take a bit more time and logistical brainpower. But they often save money (especially with rising fuel costs) and they provide additional opportunities for us to genuinely connect with others amidst

---

270   Unfortunately we live in a consumer society, so even many churches have replaced their lovely little library corners with bookshops where everything comes with ownership rights and an associated price-tag! We need to facilitate and encourage more sharing within our communities.

our busy schedules and electronically-mediated relationships. A few years ago I always made an effort to be at Richmond station (near Melbourne CBD) in time to embark on the 5:19 pm 'Card-players Express' as it had become known. Without fail a committed company of travellers would occupy at least ten (6 + 4) seats on the train, and with briefcases as tables they would play cards for the duration of the long journey down the Frankston line, until one by one they began to disembark. I am very sure that for many of them this was the highlight of their day, and another fringe benefit was that they probably looked after their work/life balance by making sure they always left work on time! Furthermore, walking and cycling help us keep fit, and mean that some no longer have to work out at the gym to exercise. Avoiding car use also helps us connect with our local neighbourhood.

We need to acknowledge that our decisions regarding where we live can make a massive difference to our carbon footprint. What trips are your most regular ones? While it is difficult to be in close proximity to everything and everyone, there is certainly wisdom in living close to where you need to be most often, as well as to good public transport connections. Such accommodation might well be more expensive, so other sacrifices might need to be made in terms of the size and age of your home. When you move home pay attention to things like a residence's orientation toward the sun, insulation, efficiency of lighting and heating, and potential capacity for the installation of solar panels on the roof.

Jet-setters Anonymous

For urban professionals (actually, for rural professionals too), it is rare for many people to think twice about the environmental impact of domestic and international air travel. More often the pressing factors being weighed up are the monetary cost of travel, and the possibility of saving 'time' when compared to driving (time is often equated to money anyway). The unhelpful practice of airlines offering cheap and easy offsets at the time of ticket purchase means that people can quite effortlessly feel satisfied that what they're doing is morally neutral. Yet flying is often the biggest contributor to a person's carbon footprint.

This is often not an easy topic to broach, as many of us have been raised with a sense of entitlement regarding air travel. As Thea Ormerod and Miriam Pepper point out in their challenging article *What is it about air travel?*, flying has become a "highly prized lifestyle option" for many of us.[271] People who work for large companies rely on regular air travel simply to make their professional lives work. Those working in cross-cultural mission, or aid and development, rely on regular air travel too. And in terms of leisure, many people have become accustomed to not only inexpensive travel, but the incredibly inexpensive holidays that can be had in overseas destinations. Want sunshine? Why go to the Sunshine Coast when you could pay half the price for an even longer holiday in Bali, within a more exotic culture thrown in? Flying with young kids is not an idea I relish, but the fact is that as a family we try to limit unnecessary flying. Accordingly, my five year old daughter is yet to experience air travel!

Like many others, we have distant friends who we would love to visit, places we want to see, things we want to do. My husband and I both immigrated to Australia as children and both have extended family overseas who we would really like to see again, while also showing our kids something of our cultural heritage. We acknowledge that there may be times in the future when flying somewhere makes sense – or is perhaps almost necessary. But in the meantime we do our best to limit flying whenever we can, and choosing holidays much closer to home is one expression of this commitment.

Make the switch / switch it off

One of the most concrete and effective steps you can take at home is to switch to 100% Government Accredited Green Power. If you haven't already done this, you can do it by picking up the phone and calling your utility provider. You might even be able to make that call right now! Go on. There is a small premium to pay, but while this is not the cheapest way to make a difference, it is certainly one of the easiest. By doing this you will ensure that the electricity

---

271 Thea Ormerod and Mirriam Pepper, "What is it about air travel?" *ARRCC*, see http://www.arrcc.org.au/what-is-it-about-air-travel, accessed 22/05/2014.

you use to run all of your lighting and electrical appliances is generated from renewable sources.[272] (It is worth noting that when Australia had a carbon pricing mechanism in place, GreenPower premiums were recalculated, making them all the more affordable!)

Australia now has solar panels installed on well over 1,000,000 roof-tops (or one in eight homes) representing one of the highest levels of market penetration in the world, and the price of systems continues to drop. That said, systems in Germany are generally twelve times larger than those on Australian roof spaces, so we are far from being global market leaders.[273] Yet rooftop solar is not an easy option for renters, or those with lots of shade or inappropriate roof-space. It is definitely worth investigating whether it is an option for your household.

Keep your eye out for new developments as the world of PV (photo-voltaic) cells is still a rapidly changing one. Battery storage is now poised to be yet another game changer as it will enable some people to remove themselves from the 'grid' entirely, eliminating the problem of power intermittency. While batteries for storing solar power remain quite costly, the price is set to decrease rapidly in the next few years as technology improves and demand increases.

Even if you purchase GreenPower and/or generate your own, the next important step is to reduce your electricity usage. It just makes sense to use less where you can, from both an environmental and financial angle. Far too many people are habitual wasters of energy. In Australia, power has traditionally been inexpensive due to our plentiful coal supply (and massive Government subsidies). As Murray Hogarth observes, "This cultural disconnect from energy as a valuable resource, and from pollution as a global threat, has led to the normalisation of wasteful behaviour in our homes and businesses."[274] This wastefulness needs to end.

272 More accurately, for every unit of electricity you use, your utility provider is obliged to buy a unit of Government Accredited Green Power. They can't guarantee that the actual electrons that flow to your house are 'green', but the industry is regulated to ensure that your equivalent usage is matched by purchases of the same amount of Green Power by your supplier.

273 Bruce Mountain, "Have solar rooftop owners had a windfall gain?" *Climate Spectator*, 20 January 2014. http://www.businessspectator.com.au/article/2014/1/20/solar-energy/have-solar-rooftop-owners-had-windfall-gain?, accessed 30/06/2014.

274 Hogarth, *The Third Degree*, 12.

If you are successful in reducing your reliance upon electricity, you will not only re-coup enough to cover the surcharge for GreenPower, but you might even save a bit of money as well – which you can use to chip away at greening your own home. This is not rocket-science. Here are some basic ideas to get you started:

- Dress appropriately for the weather to avoid unnecessary heating and cooling.[275]

- Keep doors/windows closed when using heating or air-conditioning, and seal draughts. Additionally, as much as it is safe to do so, think twice – even three times – about using your air-conditioning on hot days. Use of air-conditioning is one of the primary contributors to annual peak load/demand which is a key driver of electricity prices.[276]

- Remove incandescent light-bulbs (or halogens) and replace them with energy-saving fluorescents or LED lighting

- Turn lights and appliances off (at the wall) when they're not in use

- Only run the washing machine and dishwasher when they're full

- Think twice about whether you really need to use the clothes dryer – or take a risk and get rid of it altogether. Buy a Hills Hoist and/or an indoor clothes-airer. While you're there, turn off the second fridge, or get rid of that too.

- Have shorter showers and use low-flow shower heads: precious energy is used to heat your water (unless you take cold showers!)

- When buying new electrical appliances, check the energy rating first. Paying more for a more efficient appliance will nearly always save you money in running costs over the long-term. As always, be wary of deals that seem too good to true.

---

275 Within workplaces an acceptable range is generally considered as 20 – 24° during winter, and 23 - 26° during summer, however this takes into account that in every workforce there are generally a number of unique individuals that need to be kept happy and productive! Your thermal comfort limits at home, where you have more control, could be as low as 16° in winter or as high as 28° in summer.

276 Peak demand is a key factor driving decisions regarding new energy infrastructure, including the need for new power plants.

If you give these eight things a good go, you're already well on your way. Down the track check your utility bills (ideally the total usage, not the total price) and see how you're doing – you might be surprised at how much you can achieve if you try. Just keep in mind that for many homes, heating and cooling is the biggest energy user, so your energy use will naturally fluctuate from season to season. That's why your bill provides a comparison with 'this time last year' so you can compare apples with apples, so to speak. As for other ways to reduce your energy use and emissions, there are now many resources available online, and a reasonable list of them can be found in our resources page, as referred to at the end of this book.

It is certainly good to reduce your use of natural gas as well, as burning natural gas generates direct carbon emissions. While natural gas is not as polluting as electricity generated from burning coal, there are no genuinely green natural gas options. The best you can do is purchase an offset product (more about offsets later), but this is not ideal. Along these lines, it is probably also prudent to avoid installing new appliances that rely on natural gas. Advances in energy efficient and clean technology will focus on electrical appliances than can utilise renewable power. It has been predicted that in future many homes will shift away from gas appliances altogether (leaving fewer homes to bear the 'service charges' burden, which may result in the natural gas supply network eventually losing commercial viability).[277]

---

277 Matthew Wright, "The other demand death spiral" *Climate Spectator*, 8 November 2013. http://www.businessspectator.com.au/article/2013/11/8/energy-markets/other-demand-death-spiral, accessed 11 January 2014.

5.4 The modified mantra: Resist, Reduce, Reuse, Recycle, Repair

Key Points:

Resistance is not futile. Actually, it is essential.

Yes, you've heard it before, but here it is again: Reduce, Reuse, Recycle – the same old principles still apply.

We certainly need more trees, but be very discerning when it comes to offsets.

On the under-side of my old school bag there is a range of stickers, including one that says 'Reduce, Re-use, Recycle'. My school reunion is coming up, so I know that catchy phrase is now at least 20 years old. It's handy and memorable: a bit like the Cancer Council's *Slip, Slop, Slap*! I prefer, however, to add an extra, initial step of *resistance*, and another friend recently suggested *repair*. Daily we are bombarded with potentially hundreds of messages enticing us to consume. While the climate impact of our consumption is less direct, there are certainly emissions related to most products that we consume (through the various stages of extraction, production/manufacturing, packaging, transportation and then in some cases through some waste products at the end of an items life cycle). Indeed, a recent article challenges the way in which we so quickly point the finger at China – indeed we might find the finger is pointing right back at us:

> China is the world's largest greenhouse gas emitter, by far. The country produces more than a quarter of the planet's annual greenhouse gas emissions. World leaders increasingly reference China's spiralling emissions as a reason why it should commit to dealing with climate change. But is it fair to ask China to lead the way? After all, a hefty share of the pollution rising out of China's smokestacks comes from factories

churning out TVs, mobile phones and cheap toys for the rest of the world.[278]

In referring to this article my point is not to suggest that China's emissions are immaterial, or entirely our fault. Their national emissions make tremendous contribution to the whole, and are growing in both absolute and per capita terms. Additionally, how China responds to the pressing need to limit carbon emissions is pivotal to the success of global negotiations. That said, we need to realise that every time we purchase or consume a new product, almost inevitably there are emissions being generated somewhere in the supply chain – and often this is very far from our own shores.

As Christians we should actively seek to resist the lure and temptation of what is 'new'. We certainly need to remind ourselves that our identities are not shaped by what we have, what we use, or how we look. Character is what God values, and ultimately it is what good friends should value too. I am also convinced that God often has far more valuable uses for our finances (in fact the list of worthwhile opportunities is literally endless). But of course there is the largely unwitnessed environmental degradation that results from the extraction, manufacturing, packaging and transportation processes too. These things should also lead us to saying 'no' as, an everyday part of our discipleship.

There are two initiatives that have helped me practice the art of resistance. One is the Mutunga Partnership's $2 Challenge[279] where a household commits to live on a food and beverage budget of $2 per person, per day, for one week. Suddenly meat and treats become virtually unaffordable. Finding variety can be challenging. Planning is essential. Waste is intolerable. While the explicit aim is to save money and give away what would normally be spent to those who have less, the real value of the experience is in the learning. It is the need for resistance that many people find the most challenging: no extras, no impulse buys!

---

278 Mat Hope, "How much of China's emissions is the rest of the world responsible for?" *Climate Spectator*, 13 October, 2014. http://www.businessspectator.com.au/article/2014/10/13/science-environment/how-much-chinas-emissions-rest-world-responsible? accessed 14 October 2014.
279 "The Mutunga Challenge", http://www.mutunga.com/2dollarchallenge2.html

The other initiative is 'Buy Nothing New' month which is held in October each year.[280] Food and emergency essentials are excluded, but everything else must be made, swapped, borrowed, sourced second-hand or gone without for a whole month – including gifts for other people. The first time I participated in this program I found that it really 'broke' the lure of shops. 'Buy Nothing New' month forced me to acknowledge that I nearly always had a mental list of what I thought I 'needed' to buy next. Not necessarily expensive things. Just stuff.

As Christians, we belong to a tradition that calls followers to practise the art of resistance. What else is fasting, but learning to say 'no' to some of our everyday desires – even very legitimate ones like the food we need to sustain ourselves? While many trained in the art of fasting attest to the powerful prayer and communion with God that can accompany it, from a more practical angle avoiding cooking, eating and cleaning up after meals for any length of time frees up both time and money for other worthy pursuits – prayer, giving and serving being among the more obvious things.

The second step is *reduction*. This means less, or less often (and yes, this can mean *resistance* too)! For me, a current example is to treat meat products as much more of a treat – shifting from the everyday category to a few very small portions each week. You could reduce the length of your showers, or the frequency with which you make long journeys in your car. You could certainly try to reduce your reliance on your clothes dryer or air-conditioner. There is no end to the possibilities once you start looking to reduce. Both resisting and reducing take intentionality, choice, discipline and self-control. It is worth mentioning at this point that self-control is a fruit of the Spirit (Gal 5:22) so you're certainly not alone in the fight!

Then there is *re-using*. This one can be fun, especially when the creative juices get flowing. If you're on Facebook there are now a whole heap of sites that have some fantastic tips on re-using, re-choosing, and up-cycling. We have become

---

280 "Buy Nothing New Month", http://www.buynothingnew.com.au/ (interestingly, this initiative is supported by the Brotherhood of St Laurence, among others), accessed 20 October 2014.

a throw-away society, yet we need to keep reminding ourselves that there is no real 'away'. Things don't just disappear. The junk we throw away ends up in the ground – or worse. While throwing things out doesn't always generate direct emissions,[281] it's often a prelude to buying a brand-new replacement, the production of which has emissions implications. From a creation care perspective it is disgraceful that we so often throw out things that may still have a useful purpose. As they say, one person's trash is another's treasure!

I confess that re-using comes reasonably naturally to me – perhaps due to my frugal Scottish blood and my reasonably practical nature. I loathe waste of any kind. Accordingly I was delighted to notice a pearl of a verse in the Gospel of John: Jesus has multiplied the fish and the loaves and there is still more left after the crowds have been adequately fed. Here we meet Jesus the amazing miracle-worker who can actually make something from nothing (so one would not necessarily expect he struggles with the concept of scarcity) yet he instructs his disciples to collect what was left, saying "Let nothing be wasted". Hooray and amen! People have known about thrift since the beginning of time, but most of us have lost the art of it because we have become accustomed to such plenty, and so much of our stuff is so cheap to replace. And of course our consumption causes our economies to thrive, so we're generally encouraged to actively pursue wanton wastefulness as it makes our governments look good and is interpreted as a sure sign of our continuing economic 'prosperity'. Shopping indulgently is frequently touted as "doing something for the economy".

If you have connections with a local kinder, school or crèche, consider asking them what they need for their art and craft. You might find they are blessed to receive pieces of foam, cardboard boxes, scraps of material and other household items that we might consider junk. Again, have a think about what you are throwing out. Could you offer it on sites such as Freecycle or GumTree instead?

---

281   ... unless there is organic matter in the content which will release methane as it decomposes. But do consider the indirect emissions associated with Council vehicles transporting rubbish from your home to the tip every week or fortnight – if we all halved our waste the rubbish trucks would make 50% fewer trips to 'unload'!

Consider also making a more permanent choice to buy re-usable products rather than disposable ones. This applies to tissues (heard of hankies?), serviettes (try napkins!), nappies (yep, they're now called Modern Cloth Nappies and they don't involve pins) and even party supplies (yes, your children might only have one party each year, and Christmas is only an annual affair, but multiply each party's rubbish by the number of households in the Western world and this soon becomes a pile of waste that would probably stretch to the moon and back, and then some)!

Then there is *recycling*. Before you think about the services that your local council provides, don't forget that you can also recycle at home by composting and worm farming. Additionally, by changing your shopping habits (like avoiding polystyrene packaging trays which are not recyclable) there are now sufficient facilities and options available to enable most households to throw very little rubbish away, other than perhaps meat scraps, heavily soiled products, and foil-lined plastic packaging. Even soft packaging plastics (including cling-wrap) can be recycled at specialized depots now.[282] And, don't forget to close-the-loop by buying recycled paper products for your stationery needs, as well as things like toilet paper and paper towel. Now more than ever we need our trees left in the ground. They are our reciprocal breathing partners after all.

Finally there needs to be a commitment to *repair* what can actually be fixed. We are to be menders and makers, or "builders and fixers" rather than "wreckers and consumers" as Ben Clarke puts it.[283] Unfortunately we can usually buy new stuff so cheaply that repairing is rapidly becoming a lost art. The price signals in our consumer economy are all confused: it is cheaper to buy a new DVD player than it is to repair it, cheaper to buy new shoes than to get your favourite pair re-soled. On occasion, think about the other costs when we choose to replace instead of repair: the cost to our environment (including, for example, the precious metals contained in many technological devices, and the lives of animals where leather

---

282 Where I live in Frankston the two primary options here are facilities provided by my local Council (located at our library and the Council offices) as well as at selected Coles Supermarkets. Check out: http://www.coles.com.au/corporate-responsibility/responsible-sourcing-and-sustainability/waste, accessed 20 October 2014.

283 Ben Clarke, "A Letter to my Seven Children." *Zadok Perspectives* 117 (Summer 2012): 7.

products like shoes are involved), and also to local workers whose jobs 'fixing' things are facing extinction. Sometimes the best option isn't necessarily the cheapest option.

Speaking of trees… what about offsets?

The issue of carbon offsets could take a chapter on its own. It can get very complex, but here is a simple response:

Do your very best to live sustainably.

>Buy 100% GreenPower.

>Save energy.

>Consume less.

>Travel wisely.

Then, if you choose to offset some of your remaining emissions, make sure you do so with a very reputable organisation. I would generally recommend a not-for-profit organisation, though this doesn't always necessarily guarantee quality. Try to do some research before just ticking a box to buy an offset product, and remember that you generally get what you pay for! Many offsets are way too good to be true, particularly those sold in conjunction with purchases of airline travel:

> The [British Airways Offset] project has failed, according to one well-placed BA executive, because one part of the company wanted to improve its image by going green while another part wanted to protect its image by saying nothing at all about the impact of air travel on global warming. The result was that the scheme was launched and then banished to a dark corner of BA's website.

> That tension – between the demands of the planet and the imperatives of commerce – lies at the heart of the global response to climate change and, in particular, of carbon offsetting. The idea that we might cancel our own

greenhouse gases by paying for projects that reduce the gases elsewhere was born in the early years of climate politics. It was adopted by the corporate lobby at the Kyoto summit in 1997 and has grown into a large but deeply troubled adolescent – confused, unpredictable, and difficult to trust...

The problem with offsetting is twofold. First, these schemes are unregulated and wide open to fraud. There is nothing but the customer's canniness to stop a company claiming to be running a scheme which does not exist; claiming wildly exaggerated carbon cuts; selling offsets that have already been sold; charging hugely inflated prices... Second ... even the most well-intentioned schemes suffer from basic weaknesses in the idea of carbon offsetting – an idea which flows not from environmentalists and climate scientists trying to design a way to reverse global warming but from politicians and business executives trying to meet the demands for action while preserving the commercial status quo.[284]

Accordingly, this advice for those seeking "ethical travel" remains very relevant, even seven years on:

Instead of buying carbon offsets, try to cut out a future flight. Offsetting projects are often dubious – at their worst they involve requiring people in the Majority World not to use technology in compensation for our using it to the max. And offsets are all too often a convenient let-out for a travel industry unwilling to deal with the real implications of climate change. If you do purchase offsets, make sure you know what the company is going with the money – and don't assume it lets you off the hook of campaigning against climate change and reducing your own footprint.[285]

Increasingly, there are other kinds of offsets aside from re-forestation. You can fund renewable projects in the developing world, which doesn't reduce current emissions, but helps to ensure that the future energy needs of other communities can be met in low carbon ways.

---

284 Nick Davies, 'The inconvenient truth about the carbon offset industry' *The Guardian*, 17 Jun 2007. http://www.theguardian.com/environment/2007/jun/16/climatechange.climatechange, accessed 17 October 2014.
285 "Ten Steps to Reduce Flying." *New Internationalist* No 409, March 2008, 11.

For a number of years we have made an annual donation to Greenfleet to 'offset' the use of our vehicle, plus an extra allowance for our household landfill waste.[286] We purchase 100% GreenPower for our surplus electricity needs (i.e. beyond what we generate and use at home), and for a number of years have had an offset product bundled into our supply of natural gas.

Technically most emissions calculators would say that our household is therefore close to carbon neutral, however I am aware that every product we buy has some associated emissions – these are called *embodied emissions*.[287] So, rather than seeing the purchase of offsets as being something which makes my life magically void of negative environmental impacts, instead I see it as a willing financial contribution toward something that I know is environmentally beneficial. Funding re-forestation is an investment in natural carbon sequestration, and it can contribute to the healing of our fragile eco-systems. It is not *the* answer, but it is one solution among many. And for me this annual process is another opportunity to remember our impact on creation: it comes as a part of repentance – and there are many different 'prices' to be paid. It is also a regular reminder of the need to continue working hard toward a smaller carbon footprint as individuals, households and communities.

---

286 See "Greenfleet" website: www.greenfleet.org.au

287 Carbon emissions associated with the extraction/production and transportation processes means that there are carbon emissions embodied within most products.

5.5 Other avenues for change

Key Points:

Express your care of creation when you gather for worship.

Get busy 'greening' your other spheres of involvement (e.g. work, school).

Use your voice and your vote: we need to get 'political'.

Put your money where your values are: invest in a positive future.

Consider how you could apply yourself further, through your vocation.

Population is crucial to the discussion: we need to start the conversation.

There is another sacred cow: our diets needs serious modification.

When and where you gather

Many of the suggestions already mentioned are very applicable to places of worship. Buy 100% GreenPower. Resist, reduce, reuse, recycle and repair. Make the most of available technology to avoid distributing lots of paper-based information. Get energy aware, and use reminders and signage to ensure that those who use your facilities are up-to-speed with good switch off practices.

Here are some other helpful suggestions:

- Rather than aim for a mega-church community where people travel long distances to be involved, aim for smaller, local congregations that have deep roots within the local community.

- Ensure that sermons and teaching affirm principles of creation care, and wean people off the idea that they are "going to heaven": more correctly, the new Earth is going to be found right here and your brothers and sisters in Christ get to participate in this resurrection life along with all of creation.

- Observe 'Hope for Creation' Sunday and Earth Hour each year, and use it as a time to refocus your individual and communal energy and efforts, including a relevant sermon and helpful resources for people.

- If your church community has its own building, get the Climate Change Action Kit (from ARRCC – designed specifically for churches) and use it! Influence the Board of Elders to make changes to how energy is used in the building.

- Encourage use of real cups (not polystyrene plastic) or paper for communion and tea/coffee.

- Encourage car-pooling or use of public transport. Make community events easy to access without a car whenever possible. If your church facilities are in a public transport black-hole, lobby to get this rectified with improved transport services.

- If you have some spare land, consider starting a community garden, or really make a statement and pull up some of your concrete car park to carve out some green space.

- Support local environmental initiatives and get involved.

Monday to Friday

For those who spend much of their time at school, university or a workplace, many of the above suggestions apply, such as reducing building energy use and exploring options for more sustainable forms of travel. But you might not feel you have much capacity to bring about change yet. The first question to ask would be whether there is an environment committee, sustainability taskforce, or green team? Join them if you can, or work with one or two others to see a group established. And be the change you want to see, even if it seems small. Offer a lift, or arrange car-pooling. Explore whether grants might be available to help make the financial case for sustainability initiatives (money is often a barrier for boards/managers, but access to funding might help move things forward).

Be courageous and consistent in doing what you can to make a difference, whether it is a part of the culture of the place or not. Younger people are generally more aware and more open to such ideas, so if you are an older person in a more traditional workplace you are likely to find some solidarity and support from among the younger staff. That said, there are more and more older people who are taking climate change very seriously, so don't presume you will find resistance – step out and give it a try and you might find new friends in the most unlikely of people!

Vote + Voice

So far I have generally sought to focus on collective action, whether at a household, church or institutional level. Working together multiplies our efforts. Through building partnerships and sharing resources we can do more, and aim higher. But there are things that only you can do, such as vote. I worked as a Polling Assistant for the Australian Electoral Commission for the 2013 Federal Election and was amazed to see people throw away their votes – I was so sad to have to count fistfuls of blank ballots known as informal votes. There was so much spin, misinformation and personal attack in the lead up to the election that I think many people were all quite dismayed, confused and uninspired when it came to casting their vote. And in the midst of it all there was so little attention paid to the very pressing issue of climate change. It had been trivialised down to an argument over just another "tax" (and Julia Gillard's horrific "lie") when the focus should have been on the future of the Earth and all who will inhabit it.

I am aware that many conservative Christians in Australia lean toward voting for the Liberal Party, or another conservative or specifically 'Christian' party. While I can accept that Christians may want to support these parties for a range of reasons, when it comes to climate change, climate scientists and economists are in strong agreement that the Liberal Party's Direct Action policy is manifestly inadequate. As well as casting our votes wisely (for the House of Representatives and the Senate), as Christians we also need to make a stand and lobby our local MPs for stronger, more urgent action. If they know that sufficient people in their electorate really care, they will eventually formulate policies that adequately reflect their constituents' concern.

Elections are not the only time to make your voice heard. In our day and age there are hundreds of other ways to advocate, inform, empower, and challenge, including raising awareness through social media, supporting relevant online petitions, and getting behind other campaigns in favour of strong action on climate change. On November 17, 2013 it is estimated that more than 60,000 Australians turned out to declare their concern regarding climate change. This day included a significant number of regional events (130 in all) making it one of the nation's most widely-supported political actions. What followed in September 2014 was even bigger, and occurred on a global scale – including 400,000 people in New York City. As the saying goes, "History is made by those who show up".[288] So do all that you can to show up when it counts!

Also consider volunteering your time and energy. Contribute financially to support those working for change, if you can. And keep in mind that the powerful fossil fuel industry (and its associated lobby groups) wants action on climate change to remain voluntary: they know that people and institutions are generally reluctant to change their behaviour without being forced to – particularly if there is a cost or inconvenience involved! The aim, therefore, is to work for legislative change (yes, this involves politics) to ensure that the Australian Government supports cuts to its emissions that are equitable, measurable and significant.

---

288 This quote is generally attributed to the 19th century British Prime Minister Benjamin Disraeli.

Every nation must now step up to the plate and do their bit. Our situation is so dire that there is really no way to genuinely work for serious change without becoming politically informed and engaged. Indeed, for some (particularly those in Australia, Canada and the US) the almost unshakable influence of fossil fuel industries upon our political processes has meant a shift in strategy for climate activists. In Australia, a number of environmentally significant important policy decisions are made at State level, in particular infrastructure planning such as energy, roads and public transport. Your voice and vote matter.

Financial Assets

Globally there is a growing campaign promoting divestment from fossil fuel industries. Momentum is building, and as a result university endowments funds, local government funds, some industry-based superannuation funds and even church investment funds are being pulled out of fossil fuel-related investments. In April 2014 South African Anglican Archbishop Desmond Tutu made a public call for an anti-apartheid style boycott, calling upon people to dump investments in the fossil fuel industry, stating that "It makes no sense to invest in companies that undermine our future".[289] As McKibben articulates well "we need to take away their social license, turn them into pariahs, and make it clear that they're to the planets safety what the tobacco industry is to our individual health."[290] As at October 2014 the list of Australiasian church organisations that have made some commitment to fossil fuel divestment includes the Uniting Church Synods of NSW-ACT and Vic-Tas, at least five Anglican Dioceses in New Zealand as well as their National Anglican Synod, three Anglican Archdiocese in Australia, the New Zealand Presbyterians and National Anglican Super.

As already mentioned in Section Three, MarketForces[291] has been set up to help investors find out whether the banks and superannuation funds are investing

---

289 Damian Carrington, "Desmond Tutu calls for anti-apartheid style boycott of fossil fuel industry" *The Guardian*, 11 April 2014. http://www.theguardian.com/environment/2014/apr/10/desmond-tutu-anti-apartheid-style-boycott-fossil-fuel-industry, accessed 30 June, 2014.
290 Bill McKibben, *Oil & Honey*, 228.
291 "Market Forces" - http://www.marketforces.org.au/

funds in fossil fuels and/or the expansion of fossil fuel extraction, and to assist people in the divestment process. Now is the time to investigate how your financial assets are being utilised, and if necessary join with a divestment (boycott) action. Then, encourage others you know to do the same – including your church! Raise this issue with your priest/minister/pastor and ask them to support it if a motion comes up in synod or the equivalent (or if it's not on the radar, ask them to champion the cause themselves, and offer to support them as they do). This campaign is really gaining momentum and things are already changing as a result. While financial markets are not necessarily going to respond to the strong ethical argument that underpins divestment, the tremendous risks posed by climate change and the increasing likelihood of stranded fossil fuel assets are really beginning to sink in.[292] To demonstrate how powerful this idea is – and the threat felt by our own coal lobby groups – the battle is now being played out in the mainstream media, with some senior political leaders weighing in on the debate![293] And just as the energy retailer Powershop is changing the commercial landscape within the retail energy sector, many now hope that funds like FutureSuper will do the same in encouraging Australians to consider transferring their retirement savings into a fund that is serious about investing in an environmentally sustainable future.

While the first National Divestment Days[294] were commercial actions undertaken by banking customers, there are also a number of Christians who have engaged in non-violent direct action (otherwise known as civil disobedience). While you might not feel ready for something like this, take encouragement and inspiration from those who have been willing to take significant personal risk in stepping out to fight in this way. Encourage and support them if you come across them, as this is perhaps one of the more effective weapons we have in this season.

---

292 Nathan Fabian, "Forget divestment, all capital markets are queasy on climate", 22 October 2014, *Climate Spectator*. http://www.businessspectator.com.au/article/2014/10/22/energy-markets/forget-divestment-all-capital-markets-are-queasy-climate? accessed 25 October 2014.

293 Frank Jotzo, "Outrage at ANU divestment shows the power of its idea." *The Conversation*, 13 October 2014. http://theconversation.com/outrage-at-anu-divestment-shows-the-power-of-its-idea-32736, accessed 15/10/2014.

294 Held in May and October 2014.

Byron Smith, who was arrested at the Maules Creek Mine site in the Leard State Forest, has highlighted that some of the key fights involve questions of multi-decade infrastructure commitments – what he calls "Strategic resistance to fossil-industry infrastructure expansion". At the moment this includes, especially, new coal mines and coal ports, but also things like new motorways. If people are looking to make a meaningful contribution to the climate debate in Australia, at the moment this is where the main action is, and where we need to take our bodies and our voices.

Vocation

It is my deep hope that more and more Christians will respond to the issue of climate change as an integral expression of their vocation. Certainly not everyone needs to become a climate scientist. There will be increasing opportunity for people from a wide range of vocational disciplines to specialise in areas of environmental sustainability. As well as climatologists and meteorologists, we will need biologists (as our eco-systems change), architects (working for increasingly sustainable design), journalists (with clear, accurate reporting and a passion for the truth), teachers (especially within the sciences, politics, geography and economics), doctors (the health impacts of climate change are certainly significant), engineers (energy efficiency, construction, reconstruction, and adaptation measures), inventors (we need lots of gifted, ingenious and inspired people to solve some very serious problems), psychologists (to help us cope emotionally with life in a warming world, and with an increasingly uncertain future), entrepreneurs (looking for sustainable investment options and clever ways to move our market/s forward with good products based on sound ideas), town planners (sustainable urban design), green electricians-plumbers-tradespeople, plus a whole range of artists and communicators who can help bring this message home to stubborn humanity and help us adapt, change and grow. There are also so many new opportunities for ministry. The Church needs leaders like never before, and people who are willing to get their hands (and their green thumbs) dirty.

Lowell Bliss reflects on the Genesis 'mandate' (Genesis 1:28 and 2:15) where humankind is called to be fruitful, to fill, to subdue, to rule, to work and take care of the land, and this work is more crucial than ever:

> It's all one mandate, the essence of which is that human beings are not meant to keep themselves aloof from the creation in which they are embedded. We are to work the soil, getting it under our fingernails, partaking of its fruit, participating in its fruitfulness.[295]

As a mother, I see a crucial role for parents in raising children with a biblical worldview that is holistic, that values creation as something precious and that acknowledges our propensity to damage (often unintentionally) what God made good. So many of the decisions that are made in the home (and at the shops) will shape our children and their attitudes, and will also shape their world. Every new day provides ample opportunity for education through reflection, discussion, exploration and the asking of good questions. And of course the decision to have children at all is one of the most important vocational decisions a couple will ever make!

The (silent) population debate

Yes, the Genesis mandate that beckons us toward fruitfulness, combined with considerations of vocation, naturally brings us to a topic that most would rather avoid. While views on this topic will inevitably stir up different emotions and reactions in readers, it would be remiss to not mention the vexing issue of population at this point.

This is an extremely important and personal issue for couples to work through. It would have been easier to exclude it entirely from this conversation, not at all because it is irrelevant, but rather because it is so complex and fraught! If one wants to get into hot water with almost everyone, a safe way to do it is by starting a conversation about population. As Mick mentioned in Section One, population is a crucial factor in the global sustainability equation; there is no

---

295 Bliss, *Environmental Missions*, 64.

way around it. On one end of the spectrum there is the basic human right to procreate, and the very real need to ensure the survival of the human species. On the other end of the spectrum are the basic human rights of all people to inhabit a liveable planet with a safe climate, in the context of a planet that seems now to be filled beyond its optimal capacity. Too many people consuming too much on planet Earth jeopardises the sustainability of a whole range of global systems (weather, water, agriculture and health to name a few).

Catholic writer Paul Collins writes very succinctly and clearly on this tricky topic, summarising the difficulties on both ends of the political spectrum, and stating that the issues involved are not only difficult, but often explosive:

> Despite its importance, population is a difficult issue to discuss in polite society. Mention it and you are soon accused of being 'anti-human', an 'extreme green', 'racist', 'anti-immigrant', or wanting to dictate to developing countries about how they should behave. Or you are censured for being misinformed because the 'real issue' is not over-population, but lack of equity in the distribution of resources of the world. Perhaps the reason for these sorts of responses is that people personalise the issue, and everyone thinks their own life is of inestimable value. Consequently it's hard not to sound misanthropic when discussing population. There is also enormous vested interest in maintaining high rates of growth and immigration, especially in those Western countries which have reached zero population growth or have decreasing populations. The attempt to stifle discussion is often led by business leaders, who want to maintain the number of consumers for their goods and services without regard to the pressure this puts on the environment. In market-oriented thinking new immigrants add to the pool of consumers; they are not seen as putting strain on fragile environments. Another part of the problem is that discussion of population and the carrying capacity of particular ecosystems offends the political correctness on both the right and the left. For the right it suggests that you favour abortion, contraception, fertility control and sterilisation, especially in developing countries, and that you want to limit the rights of couples to decide the number of children they wish to have. For the left, it smacks of neo-colonialism and paternalism;

you are accused of dictating population size to developing countries and of distracting attention from social justice.[296]

Collins observes that whereas 50 years ago it was acceptable to discuss population reduction strategies as a response to environmental challenges, this is generally no longer the case. And, as a result, "political parties are often completely unwilling to discuss population issues or optimum population size for a specific nation or region and few have population policies."[297] And he goes on:

> The reason for religious people avoiding this issue is simple: it is a theological and moral minefield. It involves a whole range of acute ethical issues, as well as challenges to some ingrained attitudes within the various traditions. Perhaps the clearest of the moral conundrums we face are the ethical issues involved in intergenerational rights: if we consume so many resources now that the quality of life of future generations is compromised, are we acting in a morally responsible way?[298]

While this complex issue is worthy of consideration by all of us, it may be profoundly relevant to some readers who are facing decisions about family planning in this season of their lives. I know in my own wrestles with this, I have come across very little about this area of life and discipleship within mainstream Christian circles. This is perhaps because nobody wants to talk about it for risk of being misunderstood or of being perceived as being judgemental or self-righteous.

Certainly one thing is very clear: responsible parenting now also involves raising children who will tread lightly on God's precious Earth. This is no small feat. The pull of consumer culture is strong, and so easily grips the hearts and minds of our children, just as it does us. Equally, responsible parenting involves a commitment to work for rapid change and progress today. Time really is running out, and the lives of future generations are at stake.

---

[296] Paul Collins. *Judgment Day: The Struggle for Life on Earth*. Sydney: UNSW Press, 2010, 70.
[297] Paul Collins, *Judgment Day*, 71.
[298] Paul Collins, *Judgment Day*, 78.

The last word on another 'sacred cow'

If there is another issue as contentious and tricky as population, I think dietary restrictions might just be it! Our right to eat meat has indeed become a sacred cow. There are good biblical grounds for this: the Apostle Paul makes very clear in his letter to the church in Galatia that those who are in Christ have been set free from such laws. Not only is there now neither male nor female, nor slave or free, but the essence of his message was that there is also neither carnivore nor vegetarian!

The Jewish law was incredibly strict about dietary codes, and eating the flesh of pigs was to them an abhorrence! In light of this, Paul's message of freedom in Christ was most radical, and today's Christian leaders are rightly keen to ensure that disciples of Jesus are not weighed down by unnecessary rules or legalistic living. But Paul would also be the first to insist that we are to use our freedom for love, not for the pursuit of self-interest. Additionally his writings affirm that our use of freedom should never put a stumbling block in the path of another, particularly those who might be weaker in faith.

There are a whole range of reasons for pursuing a vegan, vegetarian or pescatarian (vegetarian + fish) diet. For many in the world these customs come naturally as a part of their religion, and a consideration of the rights of animals is at the heart of many of these customs. Others would do so for dietary reasons, with a range of diseases and medical conditions being linked to our consumption (or excessive consumption) of various animal products. Increasingly, however, more and more people are shifting toward a vegan diet for reasons of environmental sustainability. This concept is not new.

I remember being significantly impacted by Ron Sider's book *Rich Christians in an Age of Hunger* as a reasonably new Christian. Quite early on in his book Sider discusses the fact that over the life of one animal it takes seven kilograms of grain to produce each kilogram of beef (typical for what was then found in the average supermarket).[299] Not only does this mean that the rich minority of

---

299 Ron Sider, *Rich Christians in an Age of Hunger*, Dallas: Word, 1990, 23.

the world's population consume a vastly unequal proportion of the world's food, but once methane emissions are accounted for, it again puts us (as the wealthy) among those doing the most ecological damage. And it is of course not just the methane emissions themselves that contribute to climate change, but the land-clearing that is associated with cattle farming.

Additionally, methane is a far more toxic greenhouse gas than carbon dioxide. While it does not remain in the atmosphere for as long, it certainly traps a lot more heat while it is there. And to make it even worse, when compared to carbon dioxide, nitrous oxide (which is produced through the decomposition of animal urine and manure) is stronger *and* lasts longer. [For those who like numbers, over a 20 year time period carbon dioxide (CO_2) has a global warming potential (GWP) of one, Methane (CH_4) has a GWP of 56, and Nitrous Oxide (N_2O) has a GWP of 280!]

In an article by Michael Bachelard that addresses this very issue, it is noted that while deforestation for agriculture has been outlawed in Australia, clearing of regrowth is still permitted. Once this is added to general agricultural emissions (the majority of which are related to raising meat), the contribution to Australia's total emissions is almost 20 per cent![300] Accordingly, in combating climate change one of the easiest ways to reduce our national emissions would be via a wide-scale transition toward vegan diets. (While vegetarians obviously don't eat meat, the production of dairy products such as milk, cheese, yogurt and ice-cream certainly involves a whole lot of cattle.) Most Aussies would find such a proposal to be extremely confronting, to put it mildly. A transition of this kind is unlikely to be pursued as a policy platform by our politicians any time soon:

> Quite apart from the economic value of animal agriculture - $18 billion a year, including $15 billion in exports - governments are unpopular enough without invading the plates and palates of their constituents and trying to ban the barbecue.[301]

---

300 Michael Bachelard, "Some say cows are killing the earth. So do we need to ban beef?" *The Age*, 25 September, 2011. http://www.theage.com.au/environment/climate-change/some-say-cows-are-killing-the-earth-so-do-we-need-to-ban-beef-20110924-1kr2a.html, accessed 09/10/2014.

301 Michael Bachelard, "Some say cows are killing the earth. So do we need to ban beef?"

I have perhaps left this issue until last because I find it the most difficult myself. Human beings are, as they say, 'social animals'. We like to eat together. We so often celebrate with food. It is hard to make dietary choices that are not 'the norm' as they involve extra work for those cooking meals or providing hospitality, and either awkwardness or isolation – or ongoing compromise – for those who would prefer to not eat what is on offer. While the decision to become a vegan is very much a personal one, it is far easier to make this decision within the context of a family, household or broader social network that also values and supports such a decision.

If you want to start somewhere, keep in mind that it is red meat in particular, and especially beef, that has the biggest climate impact.[302] Choosing chicken over beef is something that most people can do with reasonable frequency. And while those who are vegan for reasons of animal rights are generally very uncompromising in their dietary choices (and they are not without justification, as their insight into some appalling meat industry practices are often what drives their passions and convictions), if you can't do everything it is certainly still better to do *something* rather than *nothing*. Along these lines, I have, at times, practiced a vegan brekkie/vegetarian lunch/low meat dinner pattern of eating – which is not quite as hard as one might think!

---

302 Ruminants (animals that have multiple compartments in their stomachs, including cattle and sheep) are particularly inefficient at converting food to energy and they release significantly more methane when they fart and burp. They also contribute to desertification through soil degradation. See Michael Bachelard, "Some say cows are killing the earth. So do we need to ban beef?"

Conclusion
Start by taking one step at a time

If you didn't already feel overwhelmed, you might just feel that way now! Do keep in mind that there are billions of people around the world who eat little or no meat, either out of necessity (meat is a luxury, after all) or for reasons of religious, dietary or ethical choice. These people are all making a positive contribution toward a more sustainable world, whether they even know it or not. And you can too.

Certainly the population question and the call to shift toward a vegan diet highlight yet again that there are so many ways that we must now wrestle with the pressing and urgent issue of climate change, day by day. This issue is challenging, and at times perplexing, but we must not lose hope. We must move forward toward real and helpful solutions as individuals, as workers, employers, as mums and dads, uncles and aunties, sons and daughters, grandparents – and as communities of faith. Most importantly we move forward as precious children of God. We trust immensely in his grace and provision, and the reassurance that Jesus will never leave nor forsake us. He has promised to renew all of creation, and in the meantime we are called to be people of salt and light; people of grace and truth; people known for their good works.

As an individual, it is certainly within your power to make some small, incremental changes. They will help. They may change you. And they may make others around you think again and ask some new questions: they may begin to change too. And in time you will find that you're ready to make a few more changes. Take it one step at a time, knowing that all around the world others are stepping forward just like you. You are not alone.

If we act together, it is within our power to transform the world. And this is what we have been witnessing more and more over the past 18 months. The tide of apathy, confusion and inaction really is beginning to turn.

It is our strong belief that God is calling his Church to join with those seeking action and change, for the love of God, for the love of humankind, and for the love of all that God has created.

Do you have ears to hear?

How will you respond?

Epilogue

As a layperson and self-confessed sceptic regarding the issue of climate change, reading this book has brought about a conversion, of sorts. A conversion to creation care, a call back to embracing a more holistic Gospel encompassing a care of this precious earth that God has entrusted me with.

Realising how complicit I am in contributing to a warming planet has been sobering as well as saddening. The writing is on the wall. I am able to see how I have directly contributed to climate change through my actions, my own personal greed and privatised consumption of the Earth's resources, and I can't help but feel ashamed. In all honesty, I feel helpless when faced with the sheer task of what it would take to start to repair the damage that has been done to God's Earth.

But a light has come on. Recently I finally came to understand what Mick meant when he made a contentious point during a lecture earlier this year: "False hope is heaven. True hope is earthy". While this may sound like heresy to some, it has been most helpful to me as I seek to embrace my own journey of caring for this Earth as a responsible steward of God's creation. For far too long our Christianity has subscribed to a 'pie in the sky – we'll go to heaven when we die' theology which has only absolved us of our guilty consciences as we continue to recklessly plunder God's Earth for our selfish gain.

True hope that is earthy paints a picture of getting my hands dirty, getting down on my hands and knees, firstly in lament and confession, and then literally plunging them into the soil, in my own back yard by starting to grow my own veggies (rather than gardening vicariously through my husband) together with my neighbours, many of whom live in high density flats and have no gardens! Learning how to share my resources more readily with others, dust off my bicycle seat and begin riding again, looking at ways I can join initiatives, or start my own with the help of others, so that we can reduce our collective carbon footprint on this planet, would perhaps be a good start.

As Claire and Mick have stated, caring for the Earth is a Gospel act. This is justice; this is Good News for the poor. It is justice because it is restorative, replacing and replenishing what has been taken from those who are set to suffer the worst of climate change. As affluent Australians we are certainly culpable: for a start we consume so much more than our fair share (as illustrated by the fact that we would actually need three planets if everyone on the world lived as we Aussies do)!

Jesus did far more than just die for our sins. Dying on a cross was not merely a spiritual act (even though it had significant spiritual ramifications); it involved the brutal execution of a man who stood on the side of the poor and oppressed. If we follow in Jesus' footsteps we will ourselves come against injustices such as those faced by the world's poorest and unsuspecting victims of a changing climate. The poor of this world need to see, taste, and experience this Good News through acts of restoration. This is Good News to the poor overseas and in our own street.

As a member of Urban Neighbours of Hope (a missional order called to work with the urban poor, both in Australia and Asia-Pacific), we are learning to take seriously how we can incorporate creation care in the ways we try to live out the Gospel. Good News for the poor and for us, here in Australia means offering viable alternatives that are life-giving for individuals and that are sustainable in the longer term. It is about learning to take less and give more.

Our involvement in community gardens is one small way that we have been sharing resources and promoting access to fresh, locally grown fruit and vegetables. We have been involved in initiatives such as clothing swap meets for our growing primary school children so that there are affordable alternatives for people who are on a low income and struggle to get by. Reading this book has opened up my imagination to more possibilities, yet to be fully realised and acted upon. Christianity, if we follow Jesus' example, cares for the whole person, not just their soul. This is the kind of Christianity I want to live out.

The comfort I found in reading the section on 'stories of hope' has fuelled my own hope for the future, and the future of my children, and their children. Change must begin with me. I must respond, in whatever small, seemingly insignificant way I am able, in my own neighbourhood, in the community in which I live in order to affect change. As one of my heroes, Helen Keller, said: "Alone we can do little; together we can do much". I take heart in this and the hope that together, we can make a genuine difference.

Sharmila Blair
UNOH Worker, Dandenong Victoria

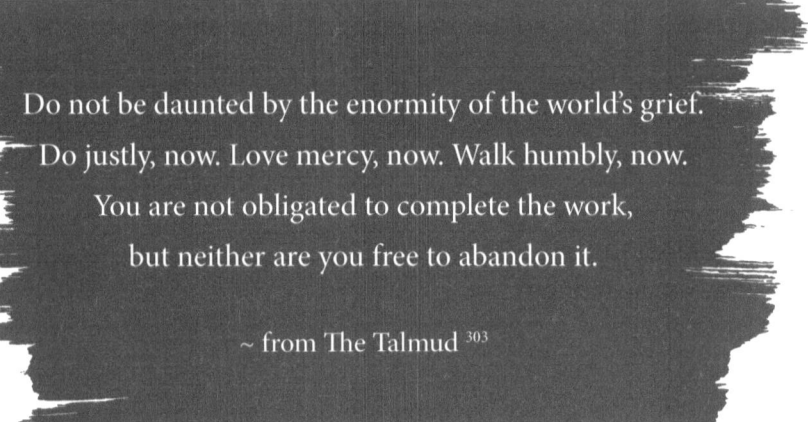

Do not be daunted by the enormity of the world's grief.
Do justly, now. Love mercy, now. Walk humbly, now.
You are not obligated to complete the work,
but neither are you free to abandon it.

~ from The Talmud [303]

---

[303] The Talmud is the central text of Rabbinic Judaism, so draws heavily upon the Old Testament.

More information?

Climate change is a moving target, and we recognise that any resource list we provided here would be out of date before it got into your hands! Instead please head to **www.aclimateofhope.com** for a list of helpful resources, including books, websites, environmental campaigns, Facebook groups and movies. We'll do our best to keep this up to date.

This book got longer than we intended. Sharmila Blair (UNOH) interviewed us back in May 2014 and we had intended that this would form the Introduction to the book. If you would like to get to know the authors a bit better, head over to our web-site and read a transcript of the interview.

Join the conversation and share your own stories of hope via the Facebook page: **www.facebook.com/climateofhope**

We again commend to you the Hope for Creation initiative and encourage you check out their web-site, sign their pledge, and follow their Facebook page: **http://hopeforcreation.com.au**

Bibliography

ABC FactCheck. "What Greg Hunt didn't say about the carbon price and emissions." *ABC News on-line*, 4 Oct 2013. http://www.abc.net.au/news/2013-10-01/greg-hunt-carbon-emissions-misleading/4989750, accessed 20 October 2014.

Anglican Church of Australia. *An Australian Prayerbook 1978*. Sydney: Anglican Information Office, 1978.

Archer, D. and Brovkin, V. "The millennial atmospheric lifetime of anthropogenic CO_2." *Climatic Change* 90 (2008): 283-297.

Archer, David. *The Long Thaw: How Humans are Changing the next 100,000 Years of Earth's Climate*. Princeton: Princeton University Press, 2009.

Bachelard, Michael. "Some say cows are killing the earth. So do we need to ban beef?" *The Age*, 25 September 2011. http://www.theage.com.au/environment/climate-change/some-say-cows-are-killing-the-earth-so-do-we-need-to-ban-beef-20110924-1kr2a.html, accessed 9 October 2014.

Balmaseda, Trenberth, and Källen, "Distinctive climate signals in reanalysis of global ocean heat content." *Geophysical Research Letters* 40 (16 May 2013): 1754-1759.

Basney, Lionel. *An Earth Careful Way of Life*, Downers Grove: IVP, 1994.

Bliss, Lowell. *Environmental Missions: Planting Churches and Trees*. Pasadena: William Carey Library, 2013.

Bouma-Prediger, Steven. *For the beauty of the Earth: A Christian vision for creation care*. 2nd edition. Grand Rapids: Baker Academic, 2010.

Carrington, Damian, 'Desmond Tutu calls for anti-apartheid style boycott of fossil fuel industry', *The Guardian*, 11 April 2014. http://www.theguardian.com/environment/2014/apr/10/desmond-tutu-anti-apartheid-style-boycott-fossil-fuel-industry, accessed 1 October 2014.

Chestney, Nina. "100 mln will due by 2030 if world fails to act on climate – report" *Reuters US Edition*, 25 September 2012. http://www.reuters.com/article/2012/09/25/climate-inaction-idINDEE88O0HH20120925, accessed 30 December 2013.

Choi, Charles. "Permian-Triassic extinction event may have been driven by extreme warming. *HuffPost Green*, 18 October 2012. http://www.huffingtonpost.com/2012/10/18/permiantriassic-extinction-event_n_1981835.html, accessed 13 February 2014.

Chubb Philip, *Power Failure: The inside story of climate politics under Rudd and Gillard*. Collingwood: Black Inc Agenda, 2014.

Clarke, Ben. "A Letter to my Seven Children" in *Zadok Perspectives* 117 (Summer 2012): 7-8.

Climate Action Centre. *4 degrees hotter: A climate action primer* (Melbourne: Climate Action Centre, 2011). http://fisnua.com/wp-content/uploads/2011/03/4-degrees-hotter.pdf, accessed 11February 2014.

Climate Council. *Off the charts: 2013 was Australia's hottest year*. Melbourne: Climate Council, 2014. http://www.climatecouncil.org.au/, accessed 3 February 2014.

Climate Institute, The. *Common Belief: Australia's faith communities on climate change*, (Sydney, December 2006). http://www.climateinstitute.org.au/articles/publications/common-belief.html, accessed 9 June 2014.

Collins, Paul. *Judgement Day: The Struggle for Life on Earth*. Sydney: UNSW Press, 2010.

Coombs, H.C. *The Return of Scarcity*. Cambridge: Cambridge University Press, 1990.

Cox, Harvey. "The market as God: living in the new dispensation", *The Atlantic Online*, 1 March 1999. http://www.theatlantic.com/magazine/archive/1999/03/the-market-as-God/306397/, accessed 6 February 2014.

Crouch, Andy in Scott McKnight and Joseph B. Modica (eds), *Jesus is Lord and Caesar is Not: Evaluating Empire in New Testament Studies*. Downers Grove: IVP Academic, 2013.

Dasgupta, Partha. and Ramanathan, Veerabhadran. "Pursuit of the Common Good." *Science*, Vol 345 No 6203, 19 September 2014. http://www.sciencemag.org/content/345/6203/1457.summary?sid=c72e5e00-16dc-4a83-a0d8-e7f67b01cc6f, accessed 5 October 2014.

Davies, Nick. "The inconvenient truth about the carbon offset industry." *The Guardian*, 17 June 2007. http://www.theguardian.com/environment/2007/jun/16/climatechange.climatechange, accessed 10/10/2014.

Dawkins, Richard. *The God Delusion*. Boston: Houghton Mifflin Harcourt, 2006.

Department of Environment. "Tuvalu's National Adaptation Programme of Action" Ministry of Natural Resources, Environment, Agriculture and Lands. Department of Environment. May 2007. unfccc.int/resource/docs/napa/tuv01.pdf, accessed 11 November, 2013.

Flinders University. "Earth Bible", http://www.flinders.edu.au/ehl/theology/ctsc/projects/earthbible/, accessed 1 November 2013.

Edis, Tristan. "Reducing pollution – not half as hard as governments think." *Climate Spectator*, 26 September 2014. http://www.businessspectator.com.au/article/2014/9/26/carbon-markets/reducing-pollution-not-half-hard-governments-think? accessed 20 October 2014.

Edwards, Julie. *The Carteret Islands: First man-made climate change evacuees still await resettlement*. Pacific Conference of Churches, November 2010. http://www.pcc.org.fj/docs/Julias%20Cartaret.pdf, accessed 11 November 2013.

Epstein, P.R. and Ferber, D. *Changing planet, changing health: How the climate crisis threatens our health and what we can do about it*. University of California: University of California Press, 2012.

Field, Christopher B. et al., *Climate Change 2014: Impacts, Adaptation, and Vulnerability, Summary for Policy Makers*. Geneva: IPCC, 2014.

Foster, Richard. *The Freedom of Simplicity*. London: Triangle, SPCK, 1981.

Friends of the Earth International, *Climate Change: Voices from Communities Affected by Climate Change*. Amsterdam: Friends of the Earth, 2007.

Galbraith, J.K. *The Affluent Society*. Mitcham: Penguin, 1958.

Garnaut, Ross. *Garnaut Climate Change Review – Update 2011, Update Paper five: The science of climate change*, Canberra: Commonwealth of Australia, 2011. http://www.garnautreview.org.au/update-2011/garnaut-review-2011.html, accessed 30 September 2014.

Gawander, J. "Impact of climate change on sugar-cane production in Fiji," *WMO Bulletin* 56:1, January 2007, 34-39.

Glikson, Andrew, "Methane and the risk of runaway global warming." *Climate Spectator*, 31 July 2013. http://www.businessspectator.com.au/article/2013/7/31/science-environment/methane-and-risk-runaway-global-warming, accessed 20 Oct 2014.

Goodman, Amy. ""If Not Now, Then When": Filipino Negotiator Pleads for Climate Deal After Typhoon Kills 500", *Democracy Now!*, 7 December 2012. http://www.democracynow.org/2012/12/7/if_not_now_then_when_filipino_negotiator, accessed 26 November 2014.

Hamilton, Clive. and Denniss, Richard. *Affluenza*. Crows Nest: Allen & Unwin, 2005.

Hamilton, Clive. *Growth Fetish*. Crows Nest: Allen & Unwin, 2003.

Hansen, J, et al. "Assessing "Dangerous Climate Change: Required reduction of carbon emissions to protect young people, future generations and nature". *PLOS One* 8 (December 2013). http://www.plosone.org/, accessed 1 October 2014.

Harper, David. "Delivering Judgement on the Great Global Warming Debate", *The Age*, 8 February 2014, http://www.theage.com.au/comment/delivering-judgment-on-the-great-global-warming-debate-20140207-32743.html, accessed 20 October 2014.

Harris, Peter. *Kingfisher's Fire: A story of hope for God's Earth*. Oxford: Monarch Books, 2009.

Hayhoe, Katharine. *A climate of change*. Nashville: Faithwords, 2009.

Hermanns, William. *Einstein and the Poet: In Search of the Cosmic Man*. Brookline Village: Branden Press, 1983.

Holmes, Jonathan. "From great moral challenge to indifference" in *Sydney Morning Herald*, 4 September 2013. http://www.smh.com.au/federal-politics/federal-election-2013/from-great-moral-challenge-to-indifference-20130903-2t355.html, accessed 30 September 2014.

Hope, Mat. "How much of China's emissions is the rest of the world responsible for?" *Climate Spectator*, 13 October 2014. http://www.businessspectator.com.au/article/2014/10/13/science-environment/how-much-chinas-emissions-rest-world-responsible? accessed 14 October 2014.

Hopkins, Keith. "Taxes and Trade in the Roman Empire (200 B.C.-A.D. 400)", *The Journal of Roman Studies*, 70 (November 1980): 101-125.

Hughes, J. Donald and Thirgood, J. V. "Deforestation, Erosion, and Forest Management in Ancient Greece and Rome", *Journal of Forest History*, 26 (April 1982): 60-75.

Jotzo, Frank. "Outrage at ANU divestment shows the power of its idea" in *The Conversation*, 13 October 2014. http://theconversation.com/outrage-at-anu-divestment-shows-the-power-of-its-idea-32736, accessed 15 October 2014.

Lansley, David. "Economic Development and World Vision: Why, how and what are we doing?" (Contained within a World Vision Report). July 2013.

Lausanne Movement, "A Confession of Faith and a Call to Action." http://www.lausanne.org/en/documents/ctcommitment.html#p1-7, accessed 20 October 2014.

Leckie, Scott. "Preparing for the climate displaced, both rich and poor" *Climate Spectator*, 24 July 2014. http://www.businessspectator.com.au/article/2014/7/24/policy-politics/preparing-climate-displaced-both-rich-and-poor? accessed 13 August, 2014.

Lenton, T. "Tipping Elements in the Earth's Climate System," *PNAS* 105 (12 February 2008): 1786-1793.

Lynas, Mark. *High Tide: How climate crisis is engulfing our planet*. New York: Harper Perennial, 2005.

Matthews, J. and Tan, H. "China roars ahead with renewable" *The Conversation*, 16 December 2013. http://theconversation.com/china-roars-ahead-with-renewables-21155, accessed 20 February 2014.

Mcbride, P. and Waterman, J. *The Colorado River: Flowing through conflict*. Englewood: Westcliffe Publishers, 2010.

McCarthy, Michael. "Lord Stern on global warming: It's even worse than I thought" *The Independent*, 13 March 2009. http://www.independent.co.uk/environment/climate-change/lord-stern-on-global-warming-its-even-worse-than-i-thought-1643957.html, accessed 20 October 2014.

McClaren, Brian. *Everything must change: Jesus, global crisis, and a revolution of hope*. Nashville: Thomas Nelson, 2007.

Mcgregor, Ian. "Australian and Canada are leading the wreckers at Warsaw." *The Conversation*, 19 November 2013. http://theconversation.com/australia-and-canada-are-leading-the-wreckers-at-warsaw-20403, accessed 20 August 2014.

McKibben, Bill, *Oil & Honey: The Education of an Unlikely Activist*. Collingwood: Black Inc., 2013.

McKibben, Bill. *Deep Economy: The Wealth of Communities and the Durable Future*. New York: Times Books, 2007.

Meckler, A.N. et al. "Deglacial pulses of deep-ocean silicate into the subtropical North Atlantic Ocean," *Nature* 495 (28 March 2013): 495-499.

Moore, C.A. "Awash in a rising sea—How global warming is overwhelming the islands of the tropical Pacific," *International Wildfire*, Jan-Feb 2002.

Mountain, Bruce. 'Have solar rooftop owners had a windfall gain?' *Climate Spectator*, 20 January 2014. http://www.businessspectator.com.au/article/2014/1/20/ solar-energy/have-solar-rooftop-owners-had-windfall-gain? accessed 30 June 2014.

Myers, Ched. *The Biblical vision of Sabbath economics*. Washington DC: Tell the Word, 2002.

New Internationalist. "Ten Steps to Reduce Flying" in *New Internationalist* 409, (March 2008): 10-11.

New, Mark and Liverman, Diana and Anderson, Kevin. "Mind the gap". *Nature Reports: Climate Change* 3 (December 2009): 143-144.

Northcott, Michael. *A moral climate: the ethics of global warming*. New York: Orbis Books, 2007.

Oreskes, N. and Conway, E. M. *Merchants of doubt*. New York: Bloomsbury Press, 2010 as well as www.merchantsofdoubt.org, accessed 20 October 2014.

Ormerod, T. & Pepper, M. 'What is it about air travel?' *ARRCC website* http://www.arrcc.org.au/what-is-it-about-air-travel, accessed 13 May 2014.

Peters, Glenn et al. "The challenge to keep global warming below 2° C." *Nature Climate Change* 3 (January 2013): 4-6.

Pope, Mick. "Preaching to the Birds? The Mission of the Church to Creation," *Speaking of Mission Volume 2*, ed. Mike Frost. Morling College: Morling College Press, 2013.

Postman, Neil, *Amusing Ourselves to Death* (2006 edition with introduction by Andrew Postman). Camberwell: Penguin, 2006.

Preston, Katherine M. "Accepting the Reality of Climate Change" *Sojourners*, December 2013. http://www.utne.com/environment/the-reality-of-climate-change-zm0z13ndzlin.aspx, accessed 10 December 2013.

Pretty, Jules. "Riches won't make you happy, but a greener economy might." *The Conversation*, 30 April 2014. http://theconversation.com/riches-wont-make-you-happy-but-a-greener-economy-might-26075, accessed 30 June 2014.

Redfern, Graham. "Australia's renewables adviser scrapes the bottom of the climate denialist barrel." *The Guardian*, 25 February 2014. http://www.theguardian.com/environment/planet-oz/2014/feb/24/climate-change-dick-warburton-sceptic-australia-renewable-energy-target-review, accessed 20 August 2014.

Responding to Climate Change. "It's time to stop this madness: Philippines plea at UN climate talks" *RTCC* 11 November 2013. http://www.rtcc.org/2013/11/11/its-time-to-stop-this-madness-philippines-plea-at-un-climate-talks/#sthash.Y5UxS2Or.dpuf, accessed 20 Oct 2014.

Richard, H. et al. "The next generation of scenarios for climate change research and assessment" *Nature* 463 (11 February 2010): 747-756.

Robine, Jean-Marie et al., "Death toll exceeded 70,000 in Europe during the summer 2003," *Comptes Rendus Biologies* 331:2 (February 2008): 171-178.

Russell, Robert John. *Cosmology, Evolution, and Resurrection Hope: Theology and Science in Creative Mutual Interaction*. Kitchener: Pandora Press, 2006.

Saddler, Hugh. 'The demand drop mystery explained' *Climate Spectator*, 6 January 2014. http://www.businessspectator.com.au/article/2014/1/6/energy-markets/demand-drop-mystery-explained, accessed 30 March 2014.

Schaeffer, Francis. *Pollution and the Death of Man: The Christian View of Ecology*. London: Hodder & Stoughton, 1970.

Seuss, Dr. and Geisel, A.S. *The Lorax*. New York: Random House, 1971.

Sider, Ron. *Rich Christians in an Age of Hunger*, Dallas: Word, 1990.

Sine, Tom. *Mustard Seed versus McWorld: Reinventing Life and Mission for a new Millennium*. London: Monarch Books, 1999.

Sine, Tom. *Wild Hope*. Kent: Monarch Publications, 1991.

Smith, John. *Advance Australia Where?* Homebush West: Anzea, 1988.

Smith, N. and Leiserowitz, A. "American evangelicals and global warming," *Global Environmental Change* 23:5 (October 2013): 1009-1017.

Snyder, Howard A. with Scandrett, Joel. *Salvation means creation healed: The ecology of sin and grace*. Eugene: Cascade Books, 2011.

Social Traders with Sustainability Victoria. *Green Social Enterprise Case Study: Green Collect*. June 1012. http://www.socialtraders.com.au/learn/dsp-default.cfm?loadref=112, accessed 28 November 2014.

St Augustine, "The literal meaning of Genesis," in *Ancient Christian Writers Volume 1*, John Hammond Taylor. New York: Paulist Press, 1982.

Stager, Curt. *Deep Future: The Next 100,000 years of Life on Earth*. Melbourne: Scribe, 2011.

Stark, Rodney. *The Rise of Christianity: How the Obscure, Marginal Jesus Movement Became the Dominant Religious Force in the Western World in a Few Centurie*. San Francisco: Harper, 1997.

Stocker, T.F et al. "Technical Summary", in *Climate Change 2013: The physical science basis. Contribution of Working Group I to the Fifth Assessment Report of the Intergovernmental Panel on Climate Change,* ed. T.F Stocker et al. Cambridge: Cambridge University Press, 2013.

Stott, John. *The Radical Disciple*. Nottingham: IVP, 2010.

Su, Reissa. "Pope Francis, Catholic Church Key to Climate Change Effort" *International Business Times*, 22 September 3014. http://au.ibtimes.com/articles/567006/20140922/pope-francis-vatican-climate-change.htm#.VDcPt9McRTs, accessed 5 October 2014.

Tainter, Joseph. *The Collapse of Complex Societies*. Cambridge University: Cambridge University Press, 1990.

Taylor, John V. *Enough is Enough*. London, SCM Press, 1975.

Toffler, Alvin. *Future Shock*. London: Pan Books, 1970.

Trenberth, K. E. and Fasullo, J. T. "An apparent hiatus in global warming?" *Earth's Future*. Doi: 10.1002/2013EF000165.

Trenberth, Kevin. "Framing the way to relate climate extremes to climate change," *Climatic Change* 115:2 (November 2012), 283-290.

Walsh, B. J. and Keesmaat, S. *Colossians remixed: Subverting the Empire*. Nottingham: IVP Academic, 2004.

Walton, John. *The lost world of Genesis One*. Nottingham: IVP, 2009.

Watts, Rikk. "The New Exodus/New Creational Restoration of the Image of God: A Biblical-Theological Perspective on Salvation" in *What does it mean to be saved: Broadening Evangelical horizons of salvation*, ed. John G. Stackhouse Jnr. Minneapolis: Baker Academic, 2002.

Welker, Michael. *Creation and Reality*. Minneapolis: Fortress Press, 2009.

Wilkinson, Loren. "Christianity and the Environment: Reflections on Rio and Au Sable", *Science and Christian Belief* (5:2, 1993).

World Health Organisation. *Climate Change and Health* Factsheet No 266, Reviewed November 2013, http://www.who.int/mediacentre/factsheets/fs266/en/, accessed 28 December 2013.

World Health Organisation. *World Malaria Report 2012 FACT SHEET,* 17 December 2012.http://www.who.int/malaria/publications/world_malaria_report_2012/wmr2012_factsheet.pdf, accessed 9 December 2013.

Wright, Matthew. "The other demand death spiral", *Climate Spectator*, 8 November 2013. http://www.businessspectator.com.au/article/2013/11/8/energy-markets/other-demand-death-spiral, accessed 11 January 2014.

Wright, N. T. *The New Testament and the People of God*. Minneapolis: Fortress Press, 1992.

Wright, N.T. *The resurrection of the Son of God*. London: SPCK, 2003.

Wright, Tom. *Paul for everyone: Galatians and Thessalonians*. Louisville: Westminster John Knox Press, 2004.

Wright, Tom. *What St Paul Really Said*. Oxford: Lion Books, 2003.

Yu, K, and Xie, S. "Recent global-warming hiatus tied to equatorial Pacific surface cooling". *Nature* 501 (28 August 2013): 403-407.

www.ingramcontent.com/pod-product-compliance
Lightning Source LLC
Chambersburg PA
CBHW032029290426
44110CB00012B/729